MW01181672

Winging It

Winging It

A Beginner's Guide to Birds of the Southwest

Catherine Coulter
Cynthia Coulter
James Coulter
and Vivian Coulter

ILLUSTRATED BY
Jennifer Owings Dewey

UNIVERSITY OF NEW MEXICO PRESS ◈ ALBUQUERQUE

10 09 08 07 06 05 04 1 2 3 4 5 6

Library of Congress Cataloging-in-Publication Data

Winging it : a beginner's guide to birds of the
Southwest / Catherine Coulter . . . [et al.] ;
illustrations by Jennifer Owings Dewey.
p. cm.
Includes index.
ISBN 0-8263-3068-1 (cloth : alk. paper)
1. Birds—Southwestern States. I. Coulter, Catherine Ann,
1947– II. Dewey, Jennifer, ill.
QL683.S75W56 2004
598'.0979—dc22
2004009386

Printed and bound in Korea by Sung-In Printing
Design and composition by Melissa Tandysh

Dedication

To R. J. Bootzin, John Crawford, Jessica Coulter,

Bessie Jensen, Frances Dunham, George M. Sutton,

our mentor and friend Jennifer Owings Dewey,

as well as all the birds of our childhood.

Table of Contents

Introduction

"Everyone is born with a bird in his heart," Frank Chapman, American ornithologist said. It must be true, because we included some birds in this book simply because we like them. Some birds are hard to find, but if you look carefully you may see a Wild Turkey, Sharp-shinned Hawk, or Sandhill Crane. Of the hundreds of kinds of birds in the southwest, it was hard for us to choose only forty-two birds for our book.

We chose many birds you are likely to see. Some birds, like Evening Grosbeaks, Steller's Jays, and House Finches, you'll see at bird feeders. Some birds you find only in forests, like the Clark's Nutcracker. We included birds that live near water, like the Wood Duck and Great Blue Heron. Some birds, like the Canyon Wren, are heard more often than seen.

As children living in the southwest, we loved being outside. We swung on tires hung from cottonwood trees. We rode our horses and fished in the pond. We built forts and fought dirt clod battles. Birds were all around us, always part of our adventures. Red-tails nested in the cottonwoods and mockingbirds sang by the kitchen door. Quail startled us when we rode across the pasture. We feel blessed to have grown up in this wonderful place.

The great bird artist, Louis Agassiz, was asked once how he had spent his vacation. He replied, "I got nearly halfway across my back yard."

Look out your window. Go outside. Listen. Birds are everywhere. They share your world too.

We hope this book helps you find the bird in your heart.

Birds of Prey

Red-tailed Hawk

Joe watched the hawk nest all spring. The nest was high in the top of a tall cottonwood tree. Now it was June. Joe couldn't stand to wait any longer. With careful moves, he slowly climbed the big tree next to the one with the hawk nest. He finally climbed high enough to spy into the nest. He saw two young hawks flapping their stubby wings.

Just then Joe heard an angry "Skreee" sound and a big hawk flew right at his face. Joe ducked. He felt the edge of the hawk's wing bap against his shoulder. The big hawk was mad! Joe scrambled to the ground as fast as he dared. Looking up, he saw both parent hawks circling in the air above him. When Joe got to the ground, he saw his brother Tim watching him.

"Gee, you came down that tree pretty fast," Tim grinned.

"It doesn't take me all day to look at a baby hawk," Joe answered.

☐ A hawk's eyesight is more powerful than a human's eyesight. Its eyes are almost as large as a full-grown man's.

☐ A Red-tailed Hawk will swoop and scream at humans who come too close to its nest.

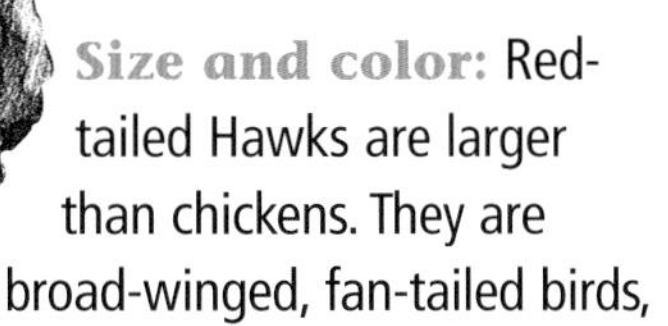

Red-tailed Hawk *(Buteo jamaicensis)*

Size and color: Red-tailed Hawks are larger than chickens. They are broad-winged, fan-tailed birds, often a reddish brown, especially the back of the tail. The shade of brown varies from region to region. Females are larger than the males. There is no red on the tails of young birds.

Habitat: all kinds of open country: woodlands, mountains, plains. They are often seen perched on fence posts and telephone poles along fields and roads.

Food: small mammals, birds, reptiles.

Nest and eggs: The nest is built with sticks and twigs high in a tree crotch, on a cliff ledge, or in a giant cactus. The nest is lined with bark, grapevine, pine needles, or moss. Redtails often use the same nesting site each year, adding a few twigs of green pine or cedar. Owls nest earlier in the year than Redtails and may take over a Redtail's nest. 2–3 creamy or bluish white eggs.

Migration: If there is plenty of food, Red-tailed Hawks stay in the same area all year. If food is scarce, they will migrate.

Call/voice: high-pitched *kreeee* or *Tseeeeeaarr.* Young hawks recently out of the nest begging for food call *Kloo-eek*.

Other names: Chicken Hawk.

Behavior:

- When you see two Redtails soaring in the same circle, you won't know if it is a courting pair. It might be two Redtails wanting the same territory. If it is two Redtails wanting to nest and hunt in the same area, one hawk will try to stay higher than the other. The highest hawk will dive down toward the lower hawk and stretch out its sharp talons. The lower hawk may flip over on its back for a moment and show its talons too. The loser of the battle will leave the area. Courting hawks do the same kind of flying acrobatics.

- Young hawks often look like they are larger than their parents because they have longer feathers. Some scientists believe the youngsters' longer feathers are necessary to make up for all the wear and tear the young birds experience as they learn to fly and hunt.

© David Ponton

The Red-tailed Hawk has a chunky body not built for speed, not like a falcon's sleek body. The Redtail is built for soaring. You'll see Redtails soaring in large circles high in the sky. You'll see Redtails sitting on fence posts or telephone poles. The Redtail is North America's most common hawk.

Peregrine Falcon

"It's spring now, Cindy. You've outgrown your jeans," Mother said. "You need some cute dresses for this warm weather."

Cindy hated shopping for clothes with her mother. She especially hated cute dresses. This will be a very long day in the city, Cindy thought. The only good thing about shopping is riding the escalators up to the top floor of the department store.

When her mother was busy talking to the sales clerk, Cindy slipped away. She knew her mother would be furious, but Cindy loved looking down on the city from the big store window. From the top floor she could see the mountains in the far distance and all the rooftops below.

© David Ponton

Most Peregrines choose to live far from people. Some, however, live in cities. The ledges of tall buildings serve as platforms for nests. City pigeons are a plentiful food supply.

Cindy was staring out at the city when a large bird flew right to the ledge outside the window. The sleek bird carried a limp pigeon in its talons. Cindy watched the big bird drop the dead pigeon on the ledge. Two fuzzy chicks waddled out from the shadows, making pitiful peeping sounds.

Cindy stepped back from the window, startled by what she had seen.

A bright red sign hung next to the window. Cindy hadn't noticed it before. The sign read: *Please be quiet. Do not tap on the glass. Peregrine Falcons nest here every year. For a close up view, watch the falcons on our web site.*

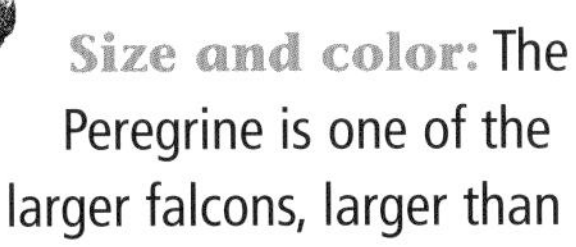

Peregrine Falcon *(Falco peregrinus)*

Size and color: The Peregrine is one of the larger falcons, larger than a pigeon. Slate gray, whitish, or buff on upper body with black bars below. It has a dark cap and a black stripe below eye.

Habitat: tundra, savannah, seacoasts, mountains, and cities with high buildings.

Food: birds such as ducks, pigeons, doves, shorebirds, and songbirds.

Nest and eggs: Peregrines nest in a hollow or scrape on the bare ground of cliff ledges or on ledges of high buildings in cities or bridges. Peregrines prefer nesting near open areas that include water. One nest site in Australia has been used by Peregrines for over 19,000 years. 2–6 eggs, buff marked with reddish or brown.

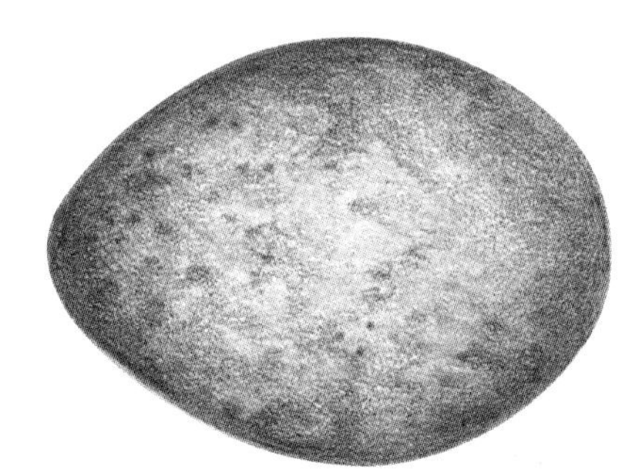

Peregrine egg.

Migration: Each Peregrine migrates alone to South America; "peregrination" means to wander or journey.

Call/voice: a long slurred cry.

Other names: Duck Hawk.

Lore: The ancient Egyptians worshipped the Peregrine. When a sacred Peregrine died, it was mummified, preserved for all eternity. In King Arthur's time, royalty trained Peregrine Falcons to hunt small animals. The falcons killed rabbits and birds and faithfully carried them back to their owners. People still practice the sport of falconry throughout the world.

Behavior:

- When a Peregrine spots its prey flying below, it folds its wings close to its body and dives. Falcons dive at speeds over 200 miles per hour. This bullet-like dive of the Peregrine is called a "stoop". The Peregrine strikes its prey with its feet, knocking it down. The Peregrine then swoops and catches the falling bird in its talons, or picks it up from the ground.

GLOSSARY

extinction: not existing, died out.

falconry: the sport of hunting with trained falcons or hawks.

prey: animal used for food.

savannah: a grassy area.

scrape: a shallow spot used for a nest, usually on the ground or a ledge.

tundra: a treeless plain of the arctic region.

Peregrine Falcon

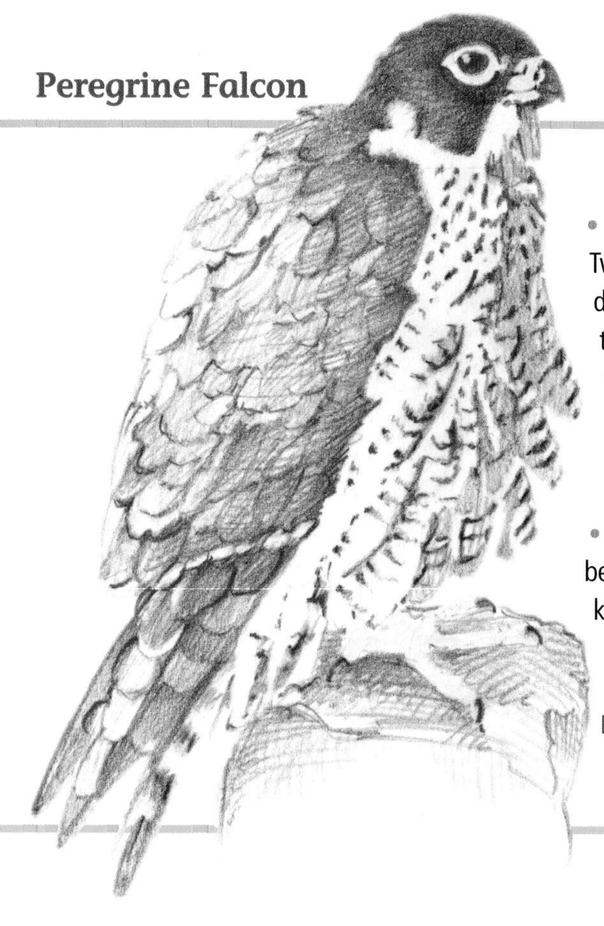

• Peregrines have a beautiful courtship flight. Two birds soar close together at high speeds, doing loop-the-loops and figure eight's. The two birds dive and chase each other across the sky. Watch closely. You might see one of the falcons roll over to face the diving mate, locking talons or bills in mid-air.

• A notch in the falcon's beak fits neatly between the neck bones of its prey for easy killing (snapping through the spinal column).

This Peregrine is fluffed up against the cold.

Scientists saved Peregrine Falcons from extinction by discovering how to hatch the eggs and raise the young in captivity. Over four thousand Peregrines have been raised from egg to fledgling and released. The Peregrine was removed from the Endangered Species list in 1999.

Today Peregrines live and nest in more than two dozen cities in the United States. Today's city-dwelling Peregrines may be captive-raised birds released from rooftops. They feel "at home" in the city and find a plentiful supply of pigeons to eat. Peregrines often return to the same nest sites year after year.

Video cameras are installed on buildings near some city nest sites. If you have a computer and a connection to the internet, search the World Wide Web and see if you can find a site that allows you to watch city-dwelling falcons and their hatchlings.

Peregrine chick.

Sharp-shinned Hawk

Among the hawks the Sharp-shinned is a veritable bushwhacker. His light body and short wings and long tail enable him to double and turn among the brush and branches, and in a noiseless, fox-like way pounce over a hedgerow or brush heap into the midst of a flock of sparrows, swoop under the low branches and pick his bird from the ground or dart through the treetops and snatch one in mid air from the midst of a startled flock.

—Florence Merriam Bailey
Handbook of Birds of the Western States (1902)

It's hard to catch a glimpse of a Sharp-shinned Hawk. The Sharp-shinned does not hunt by soaring out in the blue sky like a Redtail. This hawk lurks. It sneaks up on its prey. It flies swiftly between the branches of the forest to catch small birds.

Sharp-shinned Hawk *(Accipiter striatus)*

Size and color: somewhat larger than a robin, with a long, slim body, reddish brown chest and blue-gray head and wings, long legs. Long tail used like a rudder.

Habitat: forests, woodlands.

Food: small birds, sometimes rodents and insects.

Nest and eggs: The nest is high in a fir tree, built of branches, lined with bark. Sharp-shinned hawks are very aggressive in defense of their nest and may strike humans who come too near. 4–5 eggs, white with brown splotches.

Migration: Sharp-shinned Hawks follow the smaller birds that they prey upon.

Call/voice: *kew kew kew kew kew*, only near nest.

Other names: Sharpies: named for a sharp ridge on the leading edge of its legs.

Behavior:

- The female Sharp-shinned Hawk is much larger than the male. The female does all the brooding. The male brings food to the female three or four times a day while she incubates the eggs. She rips the prey apart for the nestlings.

GLOSSARY

brooding: raising a group of baby birds or when a parent bird keeps the young hatchlings warm by sitting over them.

bushwhacker: a person who sneaks through the woods to ambush someone.

incubating: warming the eggs while they develop.

nestlings: baby birds too little to leave the nest.

prey: animal used for food.

veritable: very much like.

- [ ] You might see a "Sharpy" raid your bird feeder in the winter. The hawk is not there for the seeds. It wants to catch small birds. Sick or weak birds are attracted to the feeder. These weak birds are the first ones caught by the hawk. Wherever you find small birds, watch for a Sharp-shinned Hawk looking for its own meal.

© Brian Wheeler/VIREO

Kestrel

"I like to watch those flag birds," Annie said, slowly eating a ripe strawberry. Annie and her mom stood in their garden, taking a break from picking strawberries.

"What's a flag bird?" asked her mother.

"Those little red, white, and blue hawks. One is on that electric line. See the red on its back and tail, blue on its wings, and white on its chest?" Annie ate two more berries. Her mother took the basket of berries. Annie watched the bird fly down and hover over a patch of weeds.

The "flag bird's" wings flapped rapidly, its tail spread wide. The bird held itself in the air for several seconds before dropping down into the weeds. A moment later the hawk flew up with a grasshopper in its talons.

"That's a kestrel," Annie's mom said. "Those little hawks look like they have eyes in the back of their head." Mom poured Annie's strawberries into her own big basket.

"Show's over. Time to get back to work." Mother handed the empty basket back to Annie.

© Tom and Pat Leeson

☐ Watch for kestrels hunting in the mornings or late afternoons. Look for a bird hovering over the ground, tail spread, beating its wings, and heading into the wind before dropping down to catch its prey. Watch the falcon return to its hunting perch, a post, or telephone wire, near an open field. After landing, the kestrel will pump its tail up and down.

American Kestrel *(Falco sparverius)*

Size and color: colorful bird about the size of a robin. It is the only *small* falcon with two "whisker" stripes on the side of its face and a rusty-red back and blue-grey wings. It has a peach colored breast with dark brown spots on its chest.

Habitat: deserts, open woodlands, grasslands, cities.

Food: Kestrels eat insects in the summer, rodents and sparrows in winter. Sometimes kestrels catch and eat lizards, scorpions, frogs, and bats.

- The kestrel has a short hooked bill and long toes with sharp talons for catching **prey**.

- Like other falcons, the female kestrel is larger than the male. If she is threatened while on her nest, she will lie on her back with her sharp **talons** raised.

Nest and eggs: in cavities in trees, saguaros, or a dirt bank. Sometimes they will nest on a cliff ledge, or in an abandoned magpie nest. Kestrels use very little nesting material. Some kestrels will use a nest box. 4–5 white or pinkish with brown eggs. They may raise a second brood if the weather and food sources are good.

Migration: Many are year round residents in the southwest. Kestrels from northern climates may winter as far south as Panama.

Call/voice: *klee klee klee* or *killy killy killy* or a *chitter* call.

Other names: Sparrow Hawk, Mouse Hawk.

Behavior:

- The kestrel has "eyes spots", two dark spots at the back of the neck. These spots look like eyes when the kestrel's head is bent over working on prey. These spots may trick other predators about which way the kestrel is facing.

- During nesting season, the male brings food to the nest area. The female leaves the nest and flies out to meet him. They both land on their perch and bow and bob their heads. She will eat some of the food and cache the rest on a branch while the male returns to hunting.

GLOSSARY

brood: group of baby birds.

cache: to store food to use later.

falcon: a type of bird of prey with long pointed wings and a hooked bill.

predator: animal who kills for food.

prey: animal used for food.

raptor: a bird of prey like a hawk.

talons: the sharp claws of a bird of prey.

Golden Eagle

Tim and Joe looked at the sky when the dark shadow passed over their heads. Each boy carried a fishing pole and a sack lunch.

"That's the biggest bird I've ever seen!" Tim said, watching a huge brown bird fly above the canyon walls.

"It's an eagle! It was carrying something in its feet," Joe said. "I think the eagle dropped something on the trail up there."

What did the eagle drop? The boys threw their poles and sacks on the ground and ran up the trail along the stream. Did the eagle drop something on purpose, or was it an accident?

"It dropped right about here," Joe said, "I'll circle out this way. You look over on that side." Tim and Joe walked slowly, searching the ground.

"Over here!" Tim shouted. Joe ran over to where Tim was standing.

"Yikes! What is that?" Joe said, "You think that's what the eagle dropped?"

"It looks like guts . . . maybe rabbit guts. It's a trick. The eagle must be nesting around here," Tim said.

"A trick? The eagle's playing a trick?" Joe asked.

"Eagles do that, drop something to lure you away from their nest," Tim said, looking up at the canyon walls, "And it worked. I don't know which way that eagle flew."

"Yum . . . rabbit guts," Joe said, "And me without my spoon!"

The boys laughed and started back along the trail to find their poles.

Size and color: Adult Golden Eagles have a six to seven foot wing spread and massive talons. The legs are feathered to the toes (The Bald Eagle, a fishing eagle, is not feathered to the feet). When seen from below, Golden Eagles are solid dark brown. There is a golden sheen on the feathers on the back of the eagle's head. Young eagles have some white on the wings and at the base of the tail.

Food: small mammals, such as ground squirrels and rabbits, large birds, carrion, lizards, snakes.

Habitat: open country, prairies, and open mountains.

☐ Golden Eagle chick.

☐ Before 1962, when a law was passed to protect eagles, some ranchers killed as many eagles as they could. The ranchers mistakenly believed they were protecting their lambs and calves. Now scientists who study eagles know that eagles rarely kill livestock. Eagles actually are helpful to ranchers by feeding on rabbits and ground squirrels.

Nest and eggs: a large platform of sticks on a cliff ledge or in a tall tree. The nest is lined with grass, leaves, or moss. 2 whitish eggs.

Migration: Eagles from the northern United States and Canada migrate to the southwest for the winter. Some eagles live in the southwest year-round.

Call/voice: not often heard.

Behavior: Eagles have problems with power lines. The tall poles are great places for eagles to perch. The problem comes when the power lines are close together. The eagle lands and its long wings touch each line. This completes an electrical circuit and the bird is electrocuted. Power companies now build poles that are safe for large perching birds.

GLOSSARY

carrion: the decaying body of a dead animal.

electrocuted: killed by electricity.

prey: animal used for food.

raptor: a bird of prey like a hawk.

- Golden Eagles kill the largest prey of any raptor: grown deer, pronghorn antelope, foxes, and coyotes. Eagles attack these large animals only if the animals are injured, when no other food is available.

Burrowing Owl

Tim walked across the prairie early one summer morning on his way to catch his horse. He heard a sound, a *bzzzzzz-bzzzzz-bzzzz*. He paused before going on, fearing a rattlesnake was near.

Tim spotted a short stubby bird, smaller than a cottontail rabbit. The bird stood on a mound of earth circling a prairie dog hole. The bird watched Tim with an unblinking stare out of big yellow eyes set in a round face.

The bird bobbed its skinny legs at the knee, bowing in Tim's direction. "Howdy to you, too," Tim said, touching the brim of his hat. "Did you make that buzzing noise?"

The noise came again, a cross between a rattle and a buzz. Tim realized the sound came from deep inside the burrow. This time Tim knew what made the noise: baby owls, out of sight underground, buzzing to scare predators away from their nest.

The opening to a Burrowing Owl nest is often lined with weed stalks and pieces of horse or cow manure. This is the owl's way of hiding its smell from predators. It uses the same materials to line the nest before laying its eggs. Heat given off by the decaying manure may help keep the eggs warm.

Size and color: Smaller than a rabbit, this little owl has sandy brown feathers mixed with white. It has black and brown bars on the wings. They have long legs and bright yellow eyes with black pupils.

Habitat: open treeless areas, mesas, prairies, parks, golf courses.

Food: crickets, grasshoppers, lizards, mice, small birds and other insects and rodents

- The Burrowing Owl's name makes you think this bird digs holes in the ground. If the ground is soft enough, the owl may dig its own burrow. Mostly though, this little owl nests in burrows of other animals. Using its long legs, beak, and wings, it enlarges a burrow left behind by a prairie dog, badger, skunk, gopher, or armadillo.

Nest and eggs: Burrowing Owls nest in colonies or family groups in abandoned burrows of rodents and reptiles. Will even nest in fields in residential areas. Sometimes digs its own burrow. 5–7 white round eggs.

Migration: They leave the northern portion of their range in the winter.

Call/voice: cooing, whistling, laughing, shrieking.

Other names: Howdy Bird, Johnny Owl, Billy Owl, Ground Owl, Badger Hole Owl, Gopher Owl, Prairie Dog Owl.

Behavior: Like other owls, this owl does not have a crop for holding its food. It swallows its prey whole, or in big chunks. Digestive juices dissolve everything except fur, feathers, and bones. The owl coughs up what is left in the form of a pellet.

- Like other owls, the Burrowing Owl has excellent vision. The eyes do not move around in the sockets. However, the owl's head moves easily side to side and up and down. And, it has long legs that help the bird peer up and over the grass around its nest, spotting food and danger.

Great Horned Owl

"I'm going to find that owl this morning!" Cindy told her mother. "I'm going to watch it hunt and see if it can turn its head all the way around." Cindy finished her oatmeal. She pulled on her sneakers and tied the strings carefully.

"Maybe," her mother said.

"Oh yes, I will! I know that old hootie is out there. I heard it again last night." Cindy lifted up her chin, "HOO! who-who-who HOO-HOO !"

Mother was surprised at Cindy's loud owl call.

"You sound just like that owl," her mother said, "And you'll surprise me again if you see an owl this morning. Owls hide during the day and leave their roosts in the evening to go off hunting. You might find the owl's roost or nest though, if you know what to look for."

"First you tell me I won't see that owl," Cindy had a stubborn look on her face. "Next you'll tell me they can't turn their heads around either?"

"Tell you what, little Miss Hootie. Run on outside. Go see and report back to me," Mother said. Cindy marched out the door.

© Tom and Pat Leeson

- Great Horned Owls have very flexible necks but they cannot swivel their heads all the way around. They can rotate their heads 180 degrees side to side. The owl's eyes have evolved to be so large that they are fixed in their sockets. That is why their necks are so flexible.

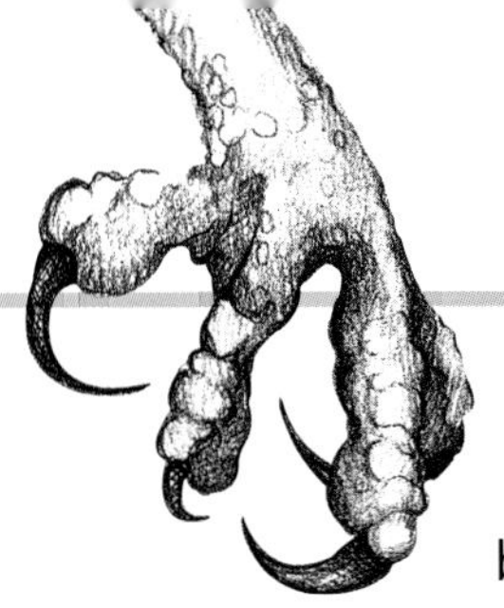

Great Horned Owl *(Bubo virginianus)*

Size and color: brown, with brown and white throat patch, 18 to 25 inches (larger than a cat). The Great Horned Owl is the only *large* owl with tufts on the head.

Habitat: forests, streamsides, open country.

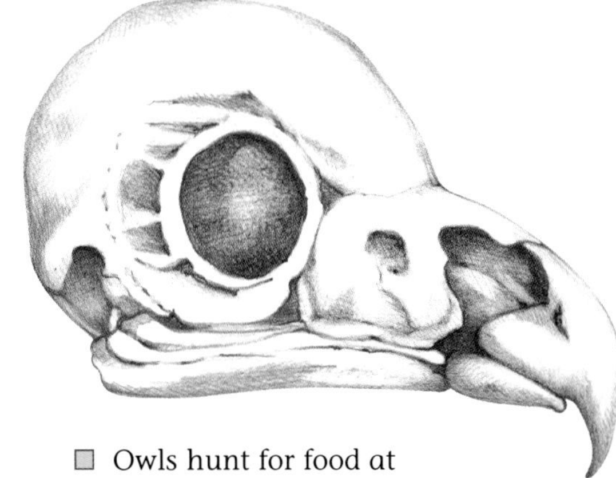

Great Horned Owl feather.

Owls hunt for food at night. Their big eyes catch enough light to see in the dark. The owl's specially designed flight feathers allow it to hunt silently.

Food: prefers rodents or rabbits. Will eat skunks, frogs, snakes, fish. Owls sometimes sit on frozen prey in the winter to thaw it before eating.

Great Horned Owls hunt by perching up high. They watch and listen for a mouse or rabbit. They swoop down and catch the prey in their talons. Sometimes an owl will catch a squirrel by slamming into the squirrel's nest. The squirrel runs out of the nest to see what happened and the owl catches it.

Great Horned Owl

- Great Horned Owls also depend upon their special ears and excellent hearing to hunt at night. The feathers around the owl's face (facial discs) act like a dish antenna to catch sound. The owl's hearing is so good it can hear a mouse moving under the snow.

Nest and eggs: uses an old hawk, crow, eagle nest. Will nest in cavities in cliffs, sometimes on the ground. 2–3 eggs.

Migration: Great Horned Owls do not migrate unless there is not enough food in their home area.

Call/voice: *HOO! who-who-who HOO-HOO!*

Other names: Hoot Owl, Hootie, Le Grand Duc (Canadian), Buho Real (Spanish), Flying Tiger.

Folklore: There are many stories about owls. Depending upon the culture, some people thought the owl wise, others thought the owl evil or linked to darkness or death.

Behavior:

• Owls eat their prey whole, swallowing bones, hair and all. Later they will regurgitate (throw-up) a pellet that contains all the undigested parts. Scientists collect these owl pellets from the ground to study the owl's diet.

• During the day, the owl is hiding. It perches with its eyes closed, body feathers flattened down and its tufts raised. This hiding way of perching (cryptic posture) helps the owl blend in with the tree trunk and bark.

GLOSSARY

cryptic posture: the way an owl hides by perching so that it blends in with the tree.

facial disc: the dish-like arrangement of feathers around an owl's face.

prey: animal used for food.

regurgitate: to vomit.

talons: the sharp claws of a bird of prey.

High Country Birds

Mountains, Forests, and Canyons

Turkey Vulture

The Turkey Vulture is the most graceful soaring bird in the world.

—Bill Kohlmoos, President of the Turkey Vulture Society

Joe and Tim lay on their backs in the grass beside the vegetable garden. They were tired of pulling weeds so they were resting in the cool grass before finishing the job.

Joe stared up at the summer sky.

"Do you see those big birds that keep circling around?" Joe asked his brother.

"Sure, I've been watching them," Tim answered. "They mostly stay above the pasture. I wonder what they're doing."

"Maybe they're vultures hunting for food. Must be something dead in that field."

"What makes you think they're vultures? Maybe they're eagles?" Tim said.

"Knucklehead! Eagles don't flock together," Joe said. "And besides, just watch their wings rock. See how their wings go tippy-tippy from side to side while they soar? That's a vulture for sure."

"Yeah, I think you're right." Tim stood up, ready to run. "I think those vultures are coming this way. I smell something dead too . . . I think it's your socks."

© David Ponton

Turkey Vultures soar and glide, holding their wings in a wide "V" shape. Vultures catch the updrafts while soaring, rocking and tipping their wings from side to side and rarely flapping.

Turkey Vulture *(Cathartes aura)*

Size and color: a large, eagle-sized black bird with a bald red head, long narrow wings with white bottom edges, deeply slotted at the tips of the wings.

Food: carrion, grass, leaves, seeds.

Nest and eggs: Vultures do not build nests. The eggs are laid on the ground on a cliff ledge, in a cave, or hollow tree. 2 eggs.

Migration: In warm areas, vultures are year-round residents. In colder areas they migrate to Central and South America for the winter, often returning to their spring roosts on the same day every year, sometimes the vernal equinox.

Call/voice: usually silent, may hiss if cornered.

Other names: The Cherokee called this bird a "peace eagle," perhaps because it does not kill.

The Vulture

The Vulture eats between his meals,
And that's the reason why
He very, very rarely feels
As well as you or I.
His eye is dull, his head is bald,
His neck is growing thinner.
Oh, what a lesson for us all
To only eat at dinner.

—Hilaire Belloc

Turkey Vulture

Behavior:

- If a Turkey Vulture is threatened or cornered, it may defend itself by projecting a smelly vomit. Or, like a possum, it may roll over and play dead.

- When vultures find a large dead animal, such as a cow, they tell other vulture flocks, about the find and lead the "guests" to the feast.

- Turkey Vultures spend several hours a day preening and cleaning themselves. They bathe in ponds where they may spend a half-hour in the water before sitting on the bank, drying their feathers.

- The Turkey Vulture's digestive system kills the viruses and bacteria in the food it eats. Its droppings and bolus (regurgitated pellets) have been studied and are found to be "clean" and do not carry disease.

Vultures are scavengers. They use their excellent eyes and keen sense of smell to find carrion. Their ability to smell carrion is especially helpful when searching over forests where the food is out of sight on the forest floor. Since the vulture is often sticking its head into a stinking carcass, it's good that its head and neck are featherless. The vulture's skin is easier to clean than feathers and the sun helps kill any bacteria on the skin.

GLOSSARY

bacteria: very small living organism.

carcass: the body of a dead animal.

carrion: the decaying body of a dead animal.

digestive system: the way the body works on food to break it down.

preening: to clean, oil, and straighten feathers.

scavenger: a bird that searches for food that is already dead.

virus: a very small organism that causes sickness.

Magpie

It is not only fine feathers that make fine birds.

—Aesop

"Jaime took my ring and won't give it back!" Cindy cried to her mother.

"What makes you think Jaime took your ring?" Mother asked.

"I laid it on the picnic table and now it isn't there. Jaime says he didn't take it," Cindy said between sobs. "Where else could it be?"

Just then a large black and white bird swooped down to the picnic table. The bird picked up a scrap of tin foil in its beak and flew away. Cindy and Mother watched the bird fly to a large nest in a tree near the creek.

"My guess is the Magpie took your ring," Mother said.

"Magpies don't wear rings," Cindy said.

"True. But they do collect shiny things. They like to hide them in their nest."

"Jaime, will you climb up there and see if my ring is in the Magpie nest?" Cindy asked her brother. "Please, please, pretty-please . . . ?"

"I'm not putting my hand in that nest. There could be a snake in there," Jaime said.

Magpies will pick up and carry shiny objects, tin foil or shards of glass, and add these to their nests. The nests are built like forts, the roofs made of stout thorn twigs. The bowls of the nest are lined with mud, rootlets, and animal hair.

Black-billed Magpie *(Pica hudsonia)*

Size and color: The flashy magpie has a black head and bill, black chest with a white belly and a long tail of shiny blue green feathers. Magpies seem like large birds because of their long tails. They are bigger than robins but smaller than ravens.

Habitat: open country with some trees; farm and ranch land and towns.

Food: insects, seeds, small animals, including dead animals, eggs and young of other birds.

Nest and eggs: large nest of sticks and twigs with a domed roof and two entrances. 6 or more eggs, greenish or buff, spotted or speckled with brown.

Migration: Magpies do not migrate.

Call/voice: can make many different loud calls but often repeats *yeck-yeck-yeck*.

Lore:

- In 1804–1806, magpies snatched food from inside the tents where members of the Lewis and Clark Expedition were eating.
- One ornithologist reported that during a time of heavy snow, a colony of magpies kept their feet warm by perching on the backs of horses and mules.

Behavior:

- The Black-billed Magpie is the naughty bird of the Bird World. This reputation is well deserved. Magpies pick at open sores on the backs of cattle and horses. They prey on other birds, snatching nestlings from nests and eating them.

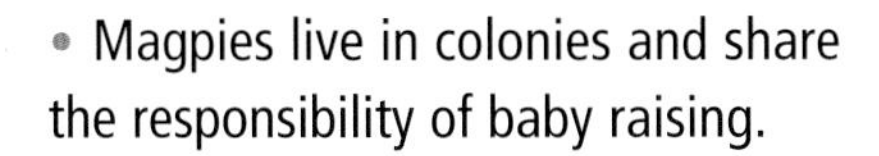

• Magpies live in colonies and share the responsibility of baby raising.

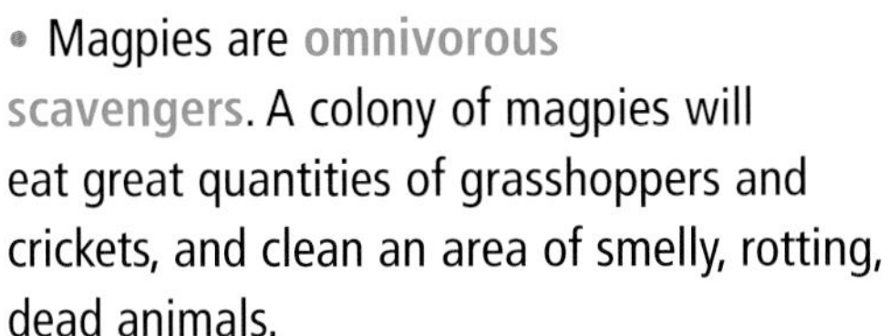

• Magpies are omnivorous scavengers. A colony of magpies will eat great quantities of grasshoppers and crickets, and clean an area of smelly, rotting, dead animals.

- Magpies are scavengers like other members of the crow family. They boldly steal food from dog bowls. They follow hunting coyotes, darting in to snatch scraps of food. Magpies do not fear bigger birds. They will scold an eagle or a falcon away from its kill. Magpies drive other birds away by flocking and dive-bombing the head of the invader.

GLOSSARY

nestling: a baby bird too young to leave the nest.

omnivorous: will eat all kinds of foods.

ornithologist: a person who studies birds.

scavenger: a bird that searches for food that is already dead.

Clark's Nutcracker

"Keep your eyes on the road," Grandpa said. Cindy was driving the tractor across the pasture. Grandpa was sitting behind her on the seat. This was Cindy's second lesson driving the tractor. Her eyes were more on the ground than on the direction they were going.

"You sure have some nice little pines trees growing in this field," Cindy said.

"Yes, right where I don't want them. I see that darn Mr. Clark has been at work again," Grandpa grumbled.

"I thought you liked Mr. Clark?" Cindy said, surprised at Grandpa's angry tone.

"I like our neighbor alright. It's *that* Mr. Clark I don't like." Grandpa pointed to a grey bird sitting on the branch of a pine tree at the edge of the pasture. Cindy stopped the tractor and watched. The large grey bird swooped to the ground near a big rock.

"Watch how he scratches his bill back and forth in the dirt," Grandpa said. "Now look how he coughs up a seed and drops it in his ditch. See how he's covering the seed with dirt?"

"That bird plants trees?"

"Sometimes the seeds sprout, but that's not the bird's purpose. Clark's nutcrackers **cache** food for winter. When that bird gets hungry this winter, he'll be back to dig up his treasures."

"How can he remember where he hid the seeds?"

"He just does . . . one smart bird, that Mr. Clark."

© David Ponton

The nutcracker doesn't eat every seed it finds. Some are stored away. The Clark's Nutcracker gathers seeds from pinecones and carries them in a pouch beneath its tongue. The pouch holds up to 90 pine seeds. A Clark's Nutcracker may **cache** over 20,000 seeds in one summer.

Clark's Nutcracker *(Nucifraga coumbiana)*

Color and size: Clark's Nutcracker is a stocky bird, bigger than a robin, with a grey head and body and a long pointed beak. It has a black tail and black wings with white patches on the tips and side.

Habitat: pine forests, high mountains.

Food: conifer seeds, insects, spiders, small animals, carrion.

Nest and Eggs: heavy well-insulated nest of twigs, grass, and pine needles in a pine tree. 2–4 pale green eggs laid in March.

Migration: Do not migrate, but they do wander during the non-breeding season.

Voice/call: short, flat caws: *caw . . . caw . . . caw*

Behavior:

- Clark's Nutcrackers bury seeds near logs or boulders and use these landmarks to find the seeds. In the winter, a hungry nutcracker uses its excellent memory and the landmarks to find the stored seeds.

- Clark's Nutcrackers eat seeds they pry out of pinecones. The bird uses its long sharp bill to jab between the pinecone scales to get to the unripe seeds.

GLOSSARY

cache: to store food to eat later.

carrion: the decaying body of a dead animal.

Steller's Jay

Cindy was playing hide-and-seek in the orchard with her brothers. But she was not happy. Big tears ran down her cheeks.

"What's the matter?" her mother called.

"The birds won't let me alone," Cindy answered. "Every time I hide, those darn jays gather around me and make a lot of noise! I'm always the first one found. It's not fair."

"Where are you hiding?" Mother asked.

"In that hedge by the swing tree," Cindy pointed.

"Sounds like the jays are mobbing you. There must be a reason," Mother said and walked out to the hedge.

Steller's Jay chick.

"Mobbing me? Does that mean attacking me? They sure sound mad." Cindy followed her mother. "What did I do to make them so mad?"

"The jays must have a nest in the hedge. They think you're a predator," Mother said. "I think you'll have to find a different place to hide."

"There it is! There's the nest!" Cindy pointed to a high place in the hedge. "Okay, Jays. I get the message. I'll find a different place."

Steller's Jay *(Cyanocitta stelleri)*

Size and color: a bit larger than a robin. The belly, wings, and tail are deep blue. They have a blackish head, crest, back, and breast.

Habitat: high mountain forests of pine or oaks, along riversides, near orchards, well-wooded suburbs.

Food: omnivorous: small vertebrates, nuts, seeds, fruits, insects, berries, sometimes eggs and young of other birds

Nest and eggs: The nest is made of sticks and twigs, lined with mud, and built in the branches of conifer trees. 3–5 greenish or blue eggs.

Migration: Most jays are year round residents, but they may move to lower elevations in winter in search of food.

GLOSSARY

cache: to store food to use later.

decoy: to lure, to trick.

habituated: used to seeing or living near.

mimic: to act like or sound like, to imitate.

omnivorous: will eat all kinds of foods.

preening: cleaning, oiling, and straightening feathers.

vertebrates: animals with backbones (mammals, birds, reptiles, amphibians, and fishes).

© David Ponton

- Steller's Jays are not frightened much by people: they are habituated to humans. Jays are often found at bird feeders and picnic tables looking for food and calling loudly.

Call/voice: loud *Shaack! Shaack!*

Behavior:

• A jay will fly far out and decoy ravens or hawks away from its nest. The jay is quiet around its nest unless threatened. A newly hatched jay is naked and featherless. This is where we get the saying "naked as a jaybird."

• Steller's Jays love acorns. They take the acorns to a high limb and hold the nut with their feet. The jay strikes the acorn with its bill to crack it. Sometimes it will cache the food to eat later.

• Steller's Jays are good mimics. They can imitate the sound of a fox squirrel, a Northern Flicker, a grosbeak, a dog, water sprinklers, and squeaky doors.

The crest is not raised when the jay is resting or preening. The jay raises its crest when it becomes angry or wants to chase other birds away from a feeder or from its nest.

Raven tracks.

Jennifer Owings Dewey and her husband adopted an orphaned raven they named Clem. He was six inches long, cold, wet, and ugly. They made him a nest in a cardboard box lined with soft fabric, fed him twelve times a day, and fell in love with him. During his first summer, Clem showed his new family just how smart ravens are.

"Clem learns to use his beak to unlatch screen doors, open the mailbox, and get into and out of the bathroom cabinet. He takes nails, screws, paintbrushes, lids to things, pieces of mail, and baby socks. The laundry hamper is one of his favorite places. Sometimes I find him wandering across the yard, dragging a piece of clothing in his beak."

—Read more in *Clem, the story of a Raven*
by Jennifer Owings Dewey

Reprint published by University of New Mexico Press, 2003.

When ravens are in flocks, one raven will sit on a high perch and watch for danger. This sentinel raven gives a loud call when it sees danger. Depending on the call, the other ravens hide or fly away. Or, the call may signal the raven flock to come together and mob a predator.

Common Raven *(Corvus corax)*

Size and color: larger than a crow, the raven is a chunky bird with blue/black feathers, diamond shaped tail, and a heavy bill.

Food: Ravens are omnivores and opportunistic scavengers. They eat carrion (that's why you see them perched along highways), crustaceans, bird eggs and young birds, insects, berries, small mammals.

Habitat: forests, deserts, urban areas.

Nest and eggs: Ravens build a large bulky nest in a tree, on cliffs, or on abandoned structures that provide support. The nest starts with sturdy sticks, then smaller sticks and twigs are added and bound together with mud, roots, and bark strips. Ravens line their nests, when possible, with fur, leaves, colored paper. 4–7 pale green eggs. The female raven moves the eggs around carefully with her beak. She turns the eggs head side up when they begin to pip (hatch).

Migration: Some ravens migrate in the fall. Some are resident, and some are partially migratory.

Raven hatchling.

GLOSSARY

cache: to store food to eat later.

carcass: the body of a dead animal.

carrion: the decaying body of a dead animal.

crustacean: a hard-shelled animal like a shrimp or crab.

mob: a group of birds chasing a predator away.

omnivore: an animal that is omnivorous (will eat all kinds of foods).

opportunistic: will take advantage of all opportunities, will eat whatever is available.

pip: break through the shell of the egg.

predator: animal who kills for food.

resident: a bird that lives in one place all year.

scavenger: a bird that searches for food that is already dead.

The raven's large bill is both a weapon against predators and a tool. The bill is strong enough to bend barbwire to include in a nest. Ravens have stiff bristle-like feathers above their bills to protect their nostrils.

Call/voice: Ravens have a large vocabulary: loud harsh territorial calls, predator alarms, rattling noises. A common call is a croaking *cr-r-uck.*

Behavior:

- Ravens will mob a Great Horned Owl, flying at it angrily, following it, and making loud calls until the owl leaves the area. The practice of mobbing is one way of teaching the young ravens in the flock which birds and animals are dangerous predators

- Ravens are probably the most intelligent of all birds. Crows and ravens use tools, solve problems, and "count" up to seven. One raven was seen to pick up rocks and drop them on the humans who were coming close to its nest.

- Ravens will follow wolves in hopes of finding a kill to feed on. Sometimes ravens lead wolves to a fresh carcass.

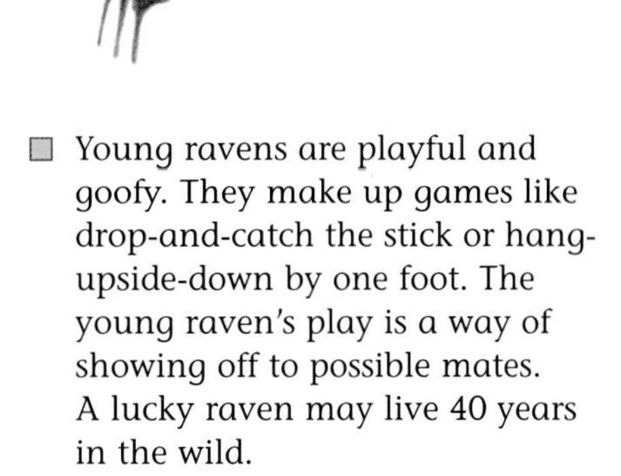

Young ravens are playful and goofy. They make up games like drop-and-catch the stick or hang-upside-down by one foot. The young raven's play is a way of showing off to possible mates. A lucky raven may live 40 years in the wild.

Cliff Swallow

True hope is swift, and flies with swallow's wings

—Shakespeare, *Richard III*

Coral watched her grandmother's hands roll red clay into long ropes. Grandmother would use the clay coils to form one of her beautiful pots. Coral tried to make her clay roll out into long, smooth, snake-shapes like her grandmother's.

"I can't do it right!" Coral said. "This takes too long."

"It does take patience. You're doing fine," Grandmother smiled.

"I wish I were a little bird," Coral said, wadding her clay into a ball. "I'd fly right out this window and swoop up to the cliffs."

"Gracious! My little girl wants to be a Cliff Swallow?" Grandmother asked. "You'll need even more patience with the clay if you're a Cliff Swallow."

"More patience?"

"Yes, you'll have to build your mud nest high on the cliff wall with the other swallows. It will take you many, many, many trips to carry the mud up to the cliff."

"Carry mud? How can I carry mud if I'm a bird?"

"In your mouth. The Cliff Swallow builds her nest slowly, one mouthful of mud at a time. It takes her a week and a thousand trips to carry all the mud she needs," Grandmother said.

"You're making me thirsty, just thinking about it." Coral took a drink of water and tried rolling her clay another time.

The swallow carries mud in its mouth to build its nest.

© David Ponton

Cliff Swallow *(Hirundo pyrrhonota)*

Size and color: a bit larger than a sparrow, with pointed wings and a square-tipped tail. Cliff Swallows are blue-black on the back, with whitish forehead, dark rusty throat and white on the belly.

Habitat: open areas, farms, cliffs, near lakes.

Food: flying insects, airborne spiders, sometimes fruit and juniper berries.

Nest and eggs: Swallows nest in colonies of a few up to a thousand nests. The nests are built under eaves of buildings, bridges or on cliff walls. The nests are made from mud pellets and are shaped like gourds with an entrance tunnel. The nest is lined with grass and feathers. 3–6 white or pinkish eggs with brown dots.

Migration: Swallows fly to South America during the winter.

call/voice: a one note call, a low *chur* or *keer*.

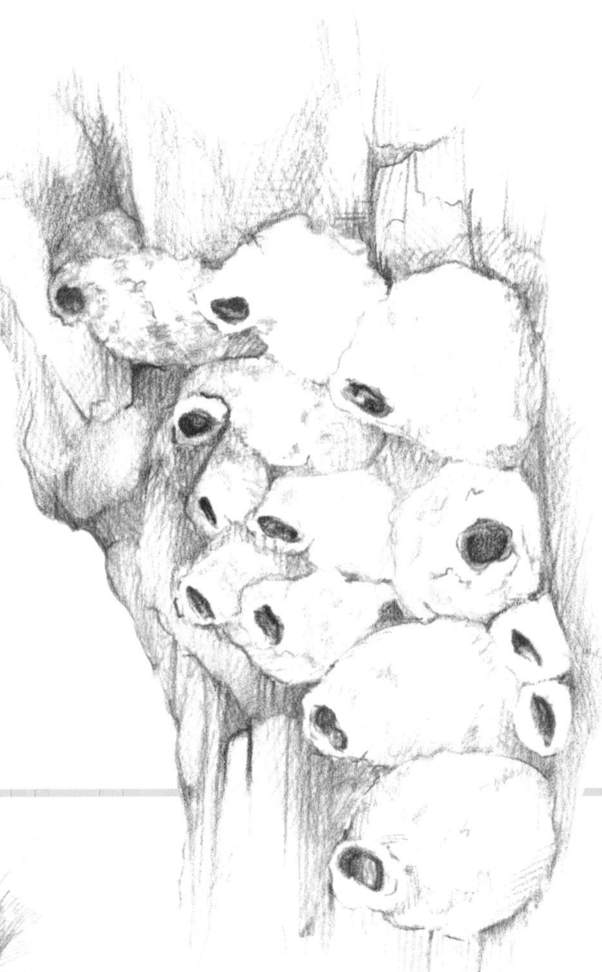

Cliff Swallows nest in colonies and reuse old nests. One colony of swallows may have as many as 1000 nests.

Cliff Swallow

Other names: Eave Swallow.

Behavior:

- The Cliff Swallow is a high-flying swallow with a wide, swooping flight. The swallow ends each glide with a steep climb.

- Although the Cliff Swallow has a small bill, it has a wide mouth (gape) for catching insects.

- The Cliff swallow flies over water and dips its bill for a quick drink while still flying. Swallows bathe the same way—by splashing up water while they fly over it.

GLOSSARY

gape: the hole made by opening the mouth wide.

© David Ponton

- Look for Cliff Swallow nests along cliff walls, barn sides, or bridges. A wall with an overhang that shields the nests is preferred. Favorite nesting places also have water, mud, and an open area for catching flying insects.

Canyon Wren

Tim and Joe carried their fishing poles over their shoulders, hiking down the path beside the stream. Their favorite trout pools were in this canyon.

"Shush! Do you hear that bird singing?" Tim whispered.

"Yes," Joe replied. "I hear it, but I don't see it anywhere. Can you see it?"

"If it's a Canyon Wren we might never see it. Dad calls them Mystery Birds because they hide from people. Maybe if we're real quiet we can hear its song again."

Tim and Joe stood still, scanning the boulders and cliff walls for any sign of movement. Soon another beautiful long whistle echoed down the canyon. The boys looked and looked, but they never caught sight of the little wren that sang the haunting song.

© Rob Curtis/VIREO

The Canyon Wren's long, thin, down-curved bill and flattened head is the perfect tool for finding insects in cracks and crevices.

Canyon Wren *(Cartherpes mexicanus)*

The Canyon Wren lives in deep canyons, along cliff sides, rock slides, boulder fields, and sometimes in stone buildings. It uses its long curved bill to find spiders, termites, ants, beetles, and other insects in the rock crevices.

Nest and eggs: The nest is built of moss, twigs, spider webs and lined with fur and feathers. It is built under rocks, or on a ledge of a cliff or in a cave. 5–6 white-brown eggs.

Migration: The Canyon Wren does not migrate. It is a permanent resident and may suffer from starvation and cold during severe winters.

Call/voice: a metallic one-note *dink*, or a longer whistled song of double notes descending in pitch, *dee-ah, dee-ah, dee-ah, dah-dah-dah*.

Behavior:

- The Canyon Wren is a shy bird that hides between rocks and in crevices. When it feels safe, the wren spends all day busily hunting insects among the rocks.

Size and color: a small, sparrow-sized brown bird with a white throat and grey head. The back and tail are spotted black and white.

Habitat: rocky canyons, cliffs, usually near flowing water.

Food: spiders, beetles, insects.

Canyon Wren hatchling.

"Go ahead, try out your new hatchet," Granddad told Jaime. "Just use it on a dead tree, not a live one."

Jaime spotted a dead pine tree near Granddad's barn. It was a cold morning and snow was on the ground, but Jaime was eager to try his new hatchet. He swung at the dead tree. The hatchet bounced off the bark without leaving a cut. Jaime kept whacking away and he heard a "conk" sound each time.

"A hollow tree . . . should be easy to cut down," Jaime said aloud as he kept chopping. Finally he found a little crack in the tree and chipped away at it until he had a hole the size of his hand. It was hard work. Jaime stopped to rest and looked up at the top of the old dead pine.

"What's that coming out of the top of the tree?" Jaime said, startled. He ran to the barn to get Granddad and begged him to come and look.

"Those are little gray birds flying out. Wow, I've never seen so many! You must have found the nuthatches' winter **roost**," Granddad said. "Let me finish the milking and I'll help you patch up the tree. The nuthatches will need their roost to make it through the winter."

- Did you ever wonder about that busy little gray bird running headfirst down the tree trunk or creeping upside-down along a branch? It's probably a noisy Pygmy Nuthatch searching for a bark beetle.

Nuthatch *(Sittidae family)*

Size and color: a small gray bird the size of your fist with a black cap and white throat, a black stripe above eye, and a very short tail.

Habitat: pine forests.

Food: weevils, leaf and bark beetles, wasps, ants; pine seeds in winter. Frequent visitor at feeders.

Nest and eggs: in a cavity, hole in a tree, a nest box. 6–8 white eggs dotted with reddish brown.

Migration: mostly year round residents. They may lower their body temperature and roost in groups to conserve energy during the winter.

Call/voice: Flocks of nuthatches keep up a steady chatter of *tidi tidi tidi*.

The White-breasted Nuthatch explores all the bark of each pine tree, poking its bill into the crevices to find insects. Nuthatches often visit bird feeders where they will fly away with a sunflower seed and wedge it tightly into a crack in the bark of a tree. It may leave it there to eat later or hammer the seed open with its bill to get at the food.

Behavior:

- Groups of nuthatches roost together in cavities in hollow trees. The nuthatches sleep in stacks to keep warm in the winter. One tree had more than 100 nuthatches roosting together. The dominant male nuthatches roost at the bottom of the cavity and the subordinate females at the top. Sometimes the birds on the bottom of the stack suffocate.

- A nuthatch covers its nest hole with the dark side of its body when threatened by an enemy. The nuthatch holds very still, making the entrance to its nest almost invisible. Or, a nuthatch may face an enemy squirrel, spread its wings, and sway from side to side. This swaying motion seems to confuse the squirrel. Soon the squirrel will give up and go away.

GLOSSARY

dominant: the one in charge of the flock, usually an older, powerful bird.

nestling: baby birds too small to leave the nest.

roost: perching to rest, often to sleep for the night, a place where birds rest.

subordinate: a younger bird that has no power or authority in the flock.

Young nuthatches often become helpers to their parents for their first year. These one-year-old males help build the nest, defend the nest, and bring food to the next batch of nestlings and the mother. Nuthatch families with young helpers raise more babies than families without helpers.

Downy Woodpecker

The woodpecker tapped his bill rapidly on the hollow tree branch. His loud drumming warned other birds to stay away. Baby woodpeckers were growing in a nest in a deep hole near the top of a nearby telephone pole. The parent woodpeckers easily ran up and down the pole bringing food to the nest.

Unlike most birds, the woodpecker has two toes pointed forward and two toes pointed back. A woodpecker's toes help the bird cling to the rough bark of a tree.

© Frank Schleicher/VIREO

Downy Woodpeckers drum on hollow wood to communicate. This sound is faster than the slow tap tap of a woodpecker pecking out a nest or looking for food. A pair of Downys will make tapping sounds to let each other know where they are.

Downy Woodpecker *(Picoides pubescens)*

Size and color: Downy Woodpeckers are the smallest black and white woodpecker. The Downy has black and white stripes on the face. The male has a red patch at the back of the head.

Habitat: forests, river groves, orchards, aspen trees. Often seen at feeders.

Food: insects and grubs, seeds and berries. They will eat suet at feeders.

Nest and eggs: The male and female dig a hole in a dead branch or tree trunk. This takes them a week. They line it with chips. Usually the nest faces south and east for full sun. 3–7 white eggs.

Migration: They stay year round, eating stored food in the winter.

Call/voice: high, soft, *whick whick* or *peek pick*.

Other names: Hewel (hew-hole).

Folklore:

- Some legends tell of woodpeckers as guardians of children. This belief may come from the fact that the woodpecker's offspring are so well hidden and protected.

Behavior:

- Downy Woodpecker nestlings sound like bees or hissing snakes. Their strange noise scares predators away from the nest.

GLOSSARY

frays: thin separated pieces.

nestlings: baby birds too small to leave the nest.

predator: animal who kills for food.

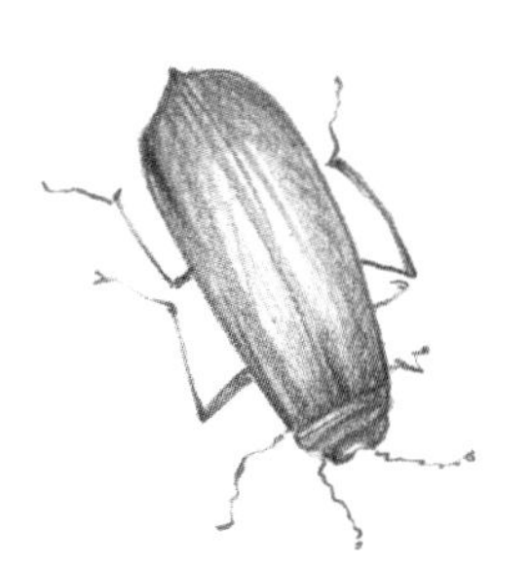

Downy Woodpecker

• The woodpecker's brain is so tightly packed into its skull that it does not rattle around and get bruised when the woodpecker drills holes in wood. The woodpecker's beak is shaped like a chisel and its long tongue is barbed and sticky: perfect for piercing cocoons and licking up insect eggs.

• Woodpeckers have a muscular pad at the base of the jaw. This pad is a shock absorber when they drum and drill into trees. The woodpecker's nostrils are protected from wood chips by little tufts of feathers at the base of the bill.

The middle of the woodpecker's tail has stiff feathers for bracing against the tree while it drills. At the end of these stiff feathers are frays that stick to the bark like Velcro and help the woodpecker stay in place.

Flicker

"What in tarnation is that awful noise?" Grandma said, holding her ears. "I'm trying to nap."

Clangity, whangity, clang!

"Sounds like Grandpa is working on the tractor," Cindy said, looking up from her book. Cindy knew naptime was quiet-time and it wasn't wise to wake Grandma.

"Impossible. Grandpa went to town," Grandma said and looked out the window. "Go tell whoever is making that awful noise to stop!"

"It's probably Joe. I'll go tell him," Cindy said, running out the door. Cindy looked around the barn but Joe was not there. She saw Joe and Tim fishing down at the pond. Cindy started back to the house.

Clang! Clang! Clang!

Cindy was so startled she jumped in the air. She turned around. Who is making that noise? Cindy saw a bird perched on the tractor seat. The bird started hammering the metal seat with its bill.

Clang! Clang! Clang!

"Shoo, you crazy woodpecker! Don't you know it's naptime?" Cindy chased the bird away. "You better not let Grandma catch you making that noise."

Cindy ran back to the house laughing. She knew the Flicker was loudly telling all the other woodpeckers he was claiming his territory for the year. She couldn't wait to tell Grandma that the Flicker was back again. On second thought, Cindy decided to wait until Grandma was up from her nap.

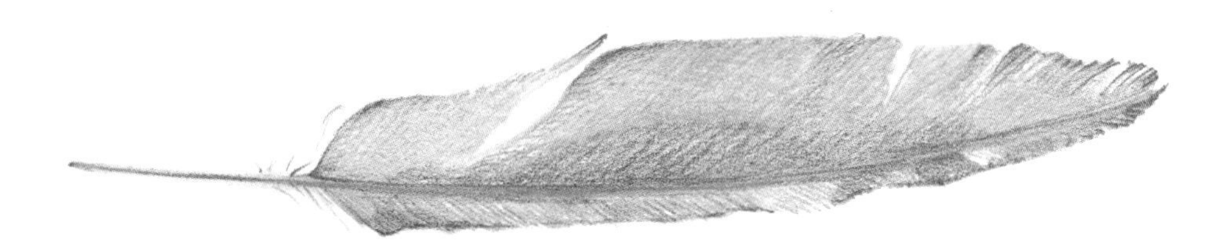

Northern Flicker *(Colaptes auratus)*

Size and color: medium large brown woodpecker, with black crescent and black spots on chest. Red stripe under eyes and red showing under wings and tail.

Habitat: riparian areas (along streams or rivers), open forests.

Food: ants and other insects, fruit and berries.

Nest and eggs: cavities in tree trunks or in large branches of dead trees, posts, saguaro cactus, or stumps. They may excavate a cavity or find one. Flickers like cavities with an entrance hole just large enough for the adult to enter. 3–8 white eggs.

Migration: Some flickers are migratory, some stay in the same area all year.

Call/voice: *wicka . . . wicka;* flickers also make drumming and tapping sounds.

© David Ponton

The most common form of Northern Flicker in the southwest is the Red-shafted Flicker. This bird's name comes from the bright red color of the feather shaft and feathers on the underside of the wings and tail.

Behavior:

- Flickers sleep straight up and down, hanging on to a tree or a wall during the night. They tuck their heads under their wings.

- The Flicker's favorite food is ants. It uses its long sticky tongue to lick the ants out of crevices.

GLOSSARY

excavate: to dig.

riparian: area near water, like streams, ponds, rivers, lakes.

- Flickers raise their babies in holes in dead trees. The hatchlings make a buzzing noise when the adult arrives at the nest hole with food. The buzzing sounds like a colony of bees and may scare enemies, like squirrels, away.

Evening Grosbeak

"Now it really feels like Thanksgiving," Mother said, laying down the nutcracker and gazing out the window.

"Is that because we're shelling pecans for pies?" Annie asked. Annie popped a handful of pecan pieces in her mouth.

"No, it feels like Thanksgiving because the Evening Grosbeaks are migrating through. See? There's a flock at the feeder right now." Mother pulled the bowl of pecan pieces from Annie's side of the table. "They always come through this time of year. I think of them as my Thanksgiving birds."

"Turkey is my Thanksgiving bird!" Annie said.

© G. Bailey/VIREO

Housecats, hawks, and owls attack **grosbeaks**. When a grosbeak spots such an enemy, it gives an alarm call. The female grosbeaks freeze where they are, while males are likely to fly for cover. Males fly away to hide their bright colors from the **predator**.

Evening Grosbeak *(Coccothraustes vespertinus)*

Size and color: Evening Grosbeaks are about the size of a robin, stocky, heavy-billed, (the greenish color of the bill is more intense in the spring). The male has a brown head shading to yellow, with white wing patches on its black wings. The female is gray overall with black tips on wings and tail.

Habitat: woodlands, forests.

Food: insects, especially budworms and other larvae, tree buds, seeds of box elder and pine, small fruits and seeds, birdfeeder seeds. Uses its big bill and powerful jaw muscles to crack seeds.

Nest and eggs: The loose cup of twigs is built far out on a limb, usually high in a pine tree. 2–5 blue green eggs with brown speckles.

Migration: many are year round residents, migrating only to lower warmer elevations in winter. Sometimes grosbeaks will come south or east (irrupt) during winter if there is a poor food supply in their home area.

Call/voice: The grosbeak rarely sings. It was named "Evening" because there was a mistaken belief this bird sang only in the evenings. It most often makes a short loud call like *cheer, keeer, peer, p-teee.*

GLOSSARY

grosbeak: large beak.

irrupt: when flocks of birds move to areas where they don't normally live.

larvae: a wingless stage of development of an insect.

predator: an animal that kills for food.

Behavior:

- An Evening Grosbeak crushes cherry pits with its strong beak. Grosbeaks drink the sap of maple trees by breaking a twig off and sipping the flowing sap.

- Evening Grosbeaks are fond of salt and will eat the salt spread on roads in the winter. This dangerous habit is why so many grosbeaks are killed by cars. Over 2000 grosbeaks were killed along one stretch of highway in British Columbia.

- A grosbeak's flimsy nest is shaped like a saucer. Despite the nest's loose structure, it holds up well. The flimsiness helps the nest blend in with the pine needles surrounding it.

Grosbeaks travel in flocks in the winter and are easy to spot at your bird feeder.

Three Things to Remember

A Robin Redbreast in a cage
Puts all Heaven in a rage.

A skylark wounded on the wing
doth make a cherub cease to sing.

He who shall hurt the little wren
Shall never be beloved by men.

—William Blake

Dark-eyed Junco *(Junco hyemalis)*

Size and color: Juncos are smaller than a robin and come in different varieties. The head and body may be blue-gray or light gray. Their eyes are circled with a dark charcoal smudge. All have white on the outer edge of the tail.

Habitat: lower foothills in the winter, mountain forests in the summer, seen on the ground under birdfeeders.

Food: seeds and insects.

Nest and eggs: a cup of grass, usually on the ground hidden under a log or bush. 3–5 whitish eggs.

Migration: small flocks migrate to the south for the winter. Some are year round residents.

Call/voice: ticking call.

Other names:
Snowbird, Mountain Junco.

Behavior:

- You see this little bird at your bird feeder in the winter, searching through seeds on the ground.
- Young male juncos spend the winter farther north than the older males. This gives the young males a head start on claiming a breeding territory in the spring.

© David Ponton

This female junco is brooding the eggs in her nest.

Mountain Bluebird

The bluebird carries the sky on its back . . .

—Thoreau

Four pink mouths gaped wide when the mother bluebird landed at the entrance to the nest box. The mother bird dropped a small insect in one baby's open mouth and flew away to find more food. The mother bird perched on a tall weed stalk in the pasture. When she spotted another bug, she swooped to the ground and caught it for her nestlings. While the mother bird hunted from one perch, the father bird hunted nearby. Both parent bluebirds hunted all day, making hundreds of trips back to the nest. The parent birds stopped to rest only after the sun went down.

Look for Mountain Bluebirds at high elevations. Mountain Bluebirds hunt for insects by perching on tall stalks in open ground. They hover low over the ground, their wings beating rapidly before dropping down to catch a grasshopper or beetle.

Mountain Bluebird *(Sialia currocoides)*

Size and color: about the size of a sparrow. The male is sky blue or turquoise blue with no reddish breast. The female is dull brown with a touch of blue on her rump, and some red on breast and sides.

Habitat: open fields with trees nearby, prairies, alpine zones above the treeline.

Food: insects, worms, snails, fruit, berries.

Nest and eggs: Bluebirds nest in holes in trees and posts, in nest boxes, dirt banks, and rock crevices; a cup of grass and pine needles, lined with feathers or animal hair. 5–6 pale blue or white eggs.

Migration: Bluebirds fly to lower elevations in the winter, such as piñon-juniper woods, farmlands, or desert.

Call/voice: a soft warbling whistle at dawn, otherwise they are silent. Navajos say bluebirds are the heralds of the rising sun.

Behavior:

- After all the baby bluebirds fledge (leave the nest), the mother will soon lay more eggs. She and the male bluebird may raise two or three families in one year.
- Sometimes the bluebird fledglings become helpers at the nest by bringing food for the next batch of babies.

GLOSSARY

fledge: to begin to leave the nest and fly.

fledgling: a young bird that is learning to fly and has left the nest but still depends on its parents for food.

herald: to call out, to announce.

hover: to hang in the air, while flapping wings.

A Mountain Bluebird nest box, part of a bluebird trail.

Bluebird Trails

Bluebirds have a hard time finding a good cavity in which to build a nest. Other birds, like sparrows and starlings, often take over the best places. To help the bluebirds, people started putting up nest boxes just for them. When people put up bluebird nest boxes along a road or trail, it is called a Bluebird Trail. These nest boxes are made to be the perfect size for bluebirds.

More bluebird trails are always needed and you can help. To find out how to build the right kind of nest box, you can write to the North American Bluebird Association at North American Bluebird Society, Inc, The Wilderness Center, P.O. Box 244, Wilmot, OH, 44689-0244 or visit their web site at www.nabluebirdsociety.org.

Mountain Chickadee

© Ron Saldino/VIREO

In the fall, chickadees cache seeds under bark, in needle clusters, and in the ground. They come back to this food in the winter.

It was dinnertime and the family gathered at the table—everyone except Cindy. Mother called her again. Quick as lightning Cindy ran to the table, snatched a piece of lettuce from the salad bowl, then ran away to the corner of the room where she poked the lettuce in her mouth. She wore a black knit ski cap and a gray sweater. The family stared at Cindy in surprised silence. She darted to the table again, grabbed a dinner roll, ran back to the corner, and started to chew on the roll.

"What in Heaven's name are you doing, Cindy?" Mother asked.

"I'm a chickadee. Joe said they grab a seed and fly off with it. That way nobody bothers them while they eat."

"Come here my little chickadee." Mother pulled Cindy onto her lap. "Does someone bother *you* while you eat?"

"Joe sticks me with his fork most days," Cindy reported.

Mountain Chickadee *(Poecile gambeli)*

Size and color: smaller than a sparrow, this handsome little bird is gray with a black cap and throat, with a white stripe above the eyes.

Habitat: pine, spruce-fir and piñon-juniper.

Food: larvae in pine needles, spruce beetles, large caterpillars, other insects, pine seeds. Frequent visitor at feeders.

Nest and Eggs: Nests are built in a cavity such as a hole in a tree or a nest box, filled with hair and wood chips. When the female leaves the nest, she covers the eggs with a hair mat. 5–7 eggs, white or white spotted with red.

Migration: Chickadees stay year round. Some young birds migrate in winter. Some birds go to lower elevations in the winter. Chickadees can go into a state of very low body temperature to conserve energy during the winter. In this slowed down sluggish state, the birds can still fly, but not very fast or far.

Call/voice: *Chick-a-dee* and *Fee-Bee-Bee.*

Behavior:

- If the nest is disturbed, the female and the nestlings make a loud hissing sound and slap the sides of the nest with their wings. This sound scares away other birds and animals.
- Chickadees have special leg muscles that make it possible for them to hang upside down at the tip of branches to get to buds and seeds.

GLOSSARY

cache: to store food to eat later.

cavities: holes in dead trees, dirt banks, rocks.

larvae: a wingless stage of development of an insect.

nestlings: baby birds too small to leave the nest

☐ Mountain Chickadees are small perky songbirds. This little bird carries food in its bill, not its feet. It flies to a perch where it holds down the seed with its feet while pounding the seed open with its bill.

Wild Turkey

"Grandma, the cow didn't come to the barn this morning," Annie said.

"Oh goodness, maybe she went off last night to have her calf." Grandma stopped stirring the pancake batter and turned off the fire under the griddle. "Animals like to be alone when they give birth. Let's see if we can find her."

Grandma wiped her hands on her apron and stepped out the screen door to the porch.

Grandma and Annie walked past the barn and followed the cow path over the pasture and into the woods. Sport bounded ahead. Sport was the farm's back-door dog and he loved any type of excursion. The little group entered the shade of the forest and stood silently for a few moments. Annie listened to the sounds in the woods. She heard her stomach rumbling and thought of the pancakes. She also heard a faint purring and scratching.

Annie watched a big turkey hen wander out of the blackberry thicket. Behind the mother turkey scurried nine baby turkeys, their black and tan fluff the same color as the oak leaves.

"Hey look, babies!" Annie shouted. The mother turkey raised her head in alarm. She let out a loud "Pulp!" sound, raised her wings, and ran straight for Sport. Sport turned and ran for home with his tail between his legs. The turkey ran with her neck stretched out after the dog. She ran almost as fast as Sport.

"What happened to all the babies?" Annie asked.

"They're right here, see?" Grandma pointed to the ground where the baby turkeys crouched motionless. She picked up a baby turkey and held it in her hand. It still didn't move. "The mother told them to freeze so they wouldn't be seen."

"We better take them home and feed them pancakes," Annie said.

"No, the turkey hen will come back. She'll tell her **poults** when it's safe to move again. She doesn't want them to get lost while she chases the danger away."

Grandma bent over and carefully put the baby turkey back on the ground. It still hadn't moved. "Let's go so the mother can come back. Besides, we have to find the cow. Then we'll go home and eat pancakes."

Wild Turkey *(Meleagris gallopavo)*

Size and color: up to four feet tall, with iridescent bronze feathers and a naked head. Turkeys have long, strong legs, a long neck, and a fan shaped tail.

Habitat: pine and oak woodlands of canyons and mountains. Turkeys need to have trees to roost in at night.

Food: acorns, wild fruits, seeds, insects.

Nest and eggs: The hen (mother turkey) scratches a little bowl on the ground among the leaves under a bush for a nest. 4–15 buff white eggs.

Migration: Wild Turkeys do not migrate.

- Turkeys make many different sounds or calls besides gobbling. Before a baby turkey hatches, it makes a peeping sound. When the mother turkey hears this sound, she makes a "yelp." As the baby turkeys grow, they make a whistling and purring sound. The mother turkey gives an "alarm putt" if it sees a predator coming or a "feeding call" to tell the flock about finding good food.

© Tom and Pat Leeson

- Wild Turkeys find most of their food, like acorns, on the ground. They scratch around in the leaves in a predictable 1-2-1 dance that hunters listen for: scratch once with one foot, twice with the other foot, once again with the first foot, then step back.

Wild Turkey

Call/voice: *gobbles, yelps, purrs, putts* depending on the situation.

Behavior:

- Special stiff bristles, called a beard, poke out from the turkey's chest. The male turkey's head has an odd kind of skin above the beak that turns red and hangs down over the beak. This special flap is called a snood. The snood and the bare skin of the turkey's head and neck are thought to help the turkey cool down when it is very hot.

- A mother turkey with very young poults (baby turkeys) keeps them warm by tucking them under her wings and tail. If a predator comes close to the turkey family, the mother gives her babies the "freeze" command and then tries to chase the predator away. The babies obey the "freeze" command by holding absolutely still until the mother returns.

- Male turkeys strut and gobble during breeding season to attract females.

GLOSSARY

Anasazi: Native American people living in the southwest before 1300 AD, ancestors of the Pueblo people.

beard: stiff feathers that poke out from the turkey's chest, usually only on the male.

domesticated: tamed by humans, raised by humans.

iridescent: shiny red, green, and purple color.

poults: young turkeys.

predator: an animal that kills for food.

snood: long skin that hangs down over the beak of a male turkey.

☐ Turkeys are a New World bird and were domesticated by the Anasazi many centuries before the arrival of the Europeans. The Anasazi people made warm blankets from the turkey's feathers. They made pictures of turkeys on the cliff walls.

Open Country Birds

Deserts, Grasslands, Backyards, Orchards

Robin

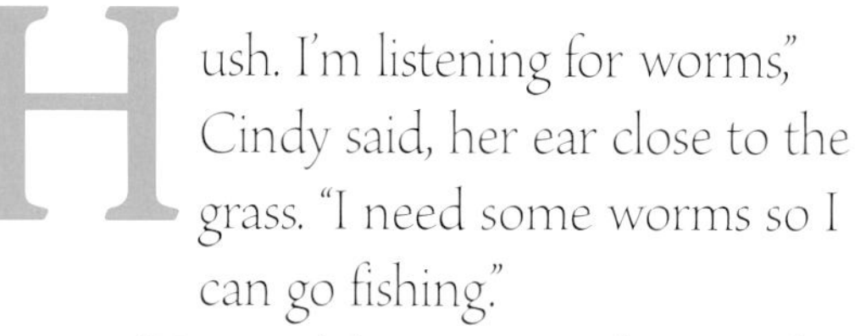

“Hush. I'm listening for worms,” Cindy said, her ear close to the grass. “I need some worms so I can go fishing.”

“You can't hear worms!” Joe said.

“Sure you can. Watch that robin. See how he hops along, then he turns his head sideways and listens?” Cindy pointed to a fat red-breasted robin patrolling the lawn. “He hears a worm and— bam! He hops on it and pulls the worm up out of its hole. I've seen him do it a zillion times.”

“The robin's not listening for worms, he's *looking for worms,*” Joe said. “See how his eyes are on the side of his head? He turns his head to the side to see the grass move.”

Cindy studied her big brother's face. Was he right or was he tricking her one more time?

This time Joe was telling Cindy the truth. After years of disagreement and debate, scientists now believe robins hunt by sight, watching for very small movements in the grass.

Call for the robin redbreast and the wren,
Since o'er shady groves they hover,
And with leaves and flowers do cover
The friendless bodies of unburied men.

—John Webster, 1612

American Robin *(Turdis migratorius)*

Size and color: 8 to 10 inches long. The male has the red breast, black head, and gray back. The female is brown and paler than the male. Young robins have speckled breasts.

Habitat: pine forests, cities, lawns, farmlands, wherever there are berries.

Food: insects, earthworms, berries

Nest and eggs: The nest is made of grass and twigs and cemented with mud. 3–6 eggs, bright "robin's egg" blue.

Migration: Robins fly to the southern United States in winter, returning north in the spring. Male robins are the first to return in the spring, followed by females, then the one-year-olds.

Call/voice: *cheerily-cheer-up-cheerio* or *be brave, be brave.*

Behavior:

- If the mother robin disappears, or dies, the father robin will take over the rearing of the young in the nest.
- The female robin carries mud to the nest in her bill. She sits in the nest and mashes the mud down into the twigs and grass with her breast.

GLOSSARY

brood: when a parent bird keeps the eggs or young hatchlings warm by sitting over them.

clutch: a group of eggs.

fledgling: a young bird that is learning to fly and has left the nest but still depends on its parents for food.

- Robin fledglings learn how to hunt for food by following the father. He also shows the young ones where to sleep at night. He leads them away from the nest and shows them where the robins roost together every evening. While the father robin is teaching the young ones, the mother robin will brood another clutch of eggs.

© David Ponton

People are fond of the robin's cheerful song, believing the song a sure sign of spring. The male robin's singing is his way of attracting a mate and claiming his territory. Sometimes you even hear a robin singing at night.

How did the robin get his red breast?

A number of cultures tell stories explaining how the male robin came by his red breast from fire or blood.

Native Americans tell how the robin brought fire to the People. On the long journey to the People, the bird had to lay its breast on the coal to keep the fire from going out.

An Irish story tells of a little bird that fanned the coals of a dying fire with his wings to keep a wolf away from a sleeping father and son. The wolf was frightened away, but the bird was left with a bright red breast.

Christians tell of a brown bird that comforted Jesus during the crucifixion, bringing him water or removing a thorn. Christ's blood fell on the bird, blessing him with a red breast ever after. Another favorite story is of the bird that came to the manger where the infant Jesus slept that first cold night. The fire in the stable was dying low and the little bird fanned the flames with his wings, bringing the fire to life.

Rufous-Sided Towhee

"Which one of you kicked my big pile of leaves apart?" Dad asked when he sat down to supper.

No one spoke. The children looked down at their plates.

"I had a big pile of leaves all raked together. I was getting ready to burn them. Now they're scattered all over the place." Father looked around the table. Cindy leaned close to Joe and whispered a question. Joe shook his head.

"Maybe it was a Doofus-Sided Toe Heel?" Cindy said.

"A what?" Dad asked.

"You know, that bird that's always scratching around in the leaves. Joe said I was acting like a Doofus-Sided Toe Heel," Cindy said, frowning at her brothers. Tim and Joe and Jaime kept looking at their plates, trying not to laugh. Bursts of giggles and snorts escaped, despite their best efforts.

"Oh . . . I think you mean a Rufous-Sided Towhee," Dad smiled, "*Now* I know who the doofus is who gets to rake the leaves up again."

Rufous-Sided Towhee* / Spotted Towhee
(Pipilo maculates)

Size and color: Smaller than a robin, the male has a black head, reddish breast and belly, and white spots on the upper part of the shoulders and wings. The female has a brown head.

Habitat: mountain thickets, piñon-juniper woods, scrub oaks.

Food: bugs, seeds, berries.

Nest and eggs: built of weed stems and grass, near or on the ground. 4–6 white eggs.

Migration: females tend to fly south during the winter while the males will often stay in one area all year or move to lower elevations for the winter.

Call/voice: *chweee*, or *shrenk* or *chup chup chup* or *drink your tea* or *zeeee*

***Other names:** Rufous-sided Towhees were recently divided into two groups: the Eastern Towhee and the Spotted Towhee found in the southwest. They were once called Chewinks, for the sound they make when surprised.

- [ ] Towhees are seen and heard on the ground where they scratch around for food. Towhees rake the ground with both feet, pushing back the leaves and digging up the dirt, ready to pounce on any seed or insect they uncover.

Oriole

Jaime heard strong winds shaking the trees during the night. He heard the rain pouring down. In the morning Jaime found a baby bird on the ground. The tiny bird peeped weakly. Its few feathers were soaked. Jaime knew exactly where its nest was. The orioles nested in the plum tree every year.

What should he do? Pick up the baby and take it in the house? Put it in the nest? Leave it alone?

Jaime ran into the house to call Dr. Sutton. Dr. Sutton was an **ornithologist**, a person who studies birds. He led Jaime's class at school on a bird-watching walk every year. He'd told Jaime's class they could call him if they had questions. Jaime had lots of questions!

"Does the baby bird have feathers?" Dr. Sutton asked.

"Sort of, but not really all over," Jaime answered.

"Then it can't fly yet. It needs to be in the nest. Pick the baby up and put it back in its nest," Dr. Sutton advised.

"But won't the mother bird think its baby smells all wrong if I pick it up?" Jaime asked.

"No, that's an old wives tale. A bird's sense of smell isn't that good," Dr. Sutton said.

"Can I put the baby bird in a box and take care of it myself?"

"No, that almost never works. The parent birds will take much better care of it than you can," Dr. Sutton said. "Since you know where the nest is, its best to put the baby back. Go quickly now, before a cat finds it."

Jaime thanked Dr. Sutton and hung up the phone. He ran outside and found the baby bird where he'd left it. Jaime put the hatchling gently in his shirt pocket and climbed the tree. The oriole parents jumped nervously around the branches of the tree as Jaime put the baby into the nest and climbed down to the ground.

Bullock's Oriole *(Icterus galbula)*

Size and color: Brightly colored and smaller than a robin, the males have black heads and orange breasts. Females are pale yellow.

Habitat: forest edges, streamsides, shade trees and cottonwood trees.

Food: bagworms, nectar, boll weevils, blueberries, grapes, other fruit.

Nest and eggs: A well-woven bag or pouch suspended from the tip of a branch. 4–6 grayish spotted eggs.

Migration: Orioles migrate to southern Mexico, Central America, and Colombia in late summer.

Call/voice: piping whistled notes.

Other names: Firehang Bird, Golden Robin.

GLOSSARY

larva: an insect in the wingless stage of development.

ornithologist: a person who studies birds.

- ☐ Orioles will come to a nectar feeder, unlike most other birds. Sometimes they learn to sip from a hummingbird feeder. Special oriole feeders hold oranges.

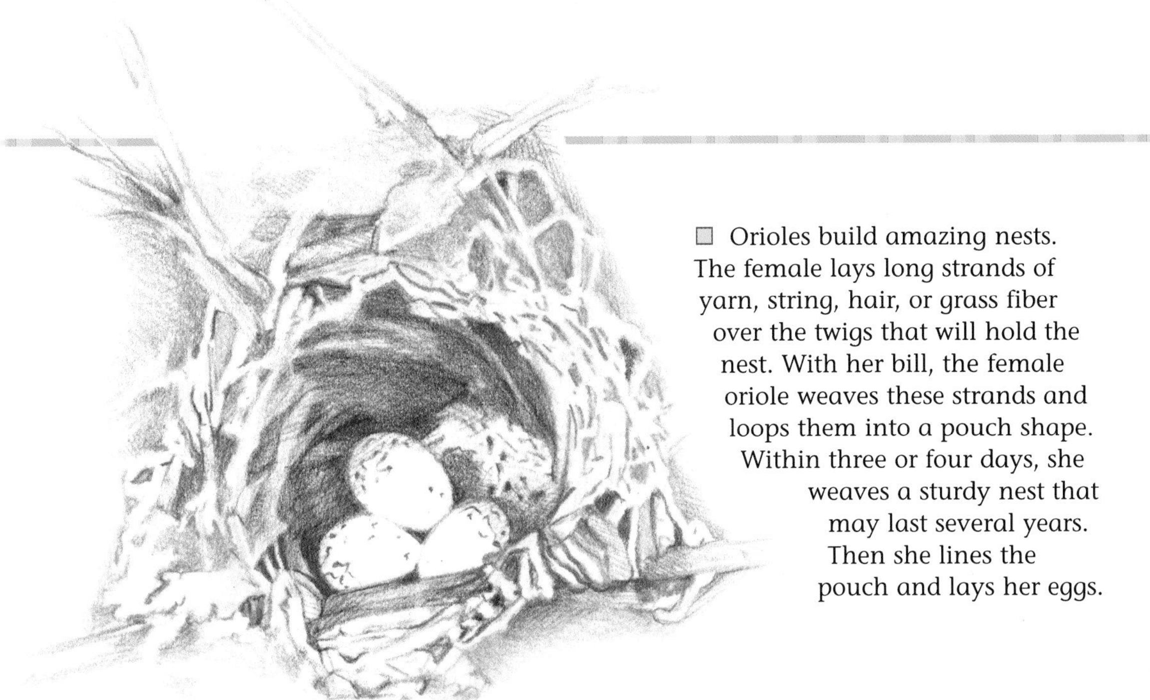

□ Orioles build amazing nests. The female lays long strands of yarn, string, hair, or grass fiber over the twigs that will hold the nest. With her bill, the female oriole weaves these strands and loops them into a pouch shape. Within three or four days, she weaves a sturdy nest that may last several years. Then she lines the pouch and lays her eggs.

Behavior:

- Orioles will use moss, hair, and plant fibers to make the hanging pear-shaped nest. If yarn and string are available, orioles always pick drab colors for their nests. If there is a choice of yarn colors, orioles pick the white and pale grey pieces.
- The oriole's beak is strong, straight, and pointed. The oriole uses its beak like a needle to weave the fibers and strands of the nest.
- Look for orioles in the tops of shade trees. Elm trees and orchard trees are favorites for orioles.

□ The oriole sneaks gently down a branch to catch a bagworm. The oriole knows if it shakes the branch, the bagworm larvae will hide deep in the indestructible bag. The oriole sneaks up, waits for the larva to poke its head out, grasps it, and crushes the larva.

Mockingbird

Mockingbirds don't do one thing but make music for us to enjoy. They don't eat up people's gardens, don't nest in corncribs, they don't do one thing but sing their hearts out for us. That's why it's a sin to kill a mockingbird.

—Harper Lee, *To Kill a Mockingbird,* 1960.

Every spring a pair of mockingbirds nested in the plum tree outside the kitchen door. Every spring the birds attacked Joe and his brothers and sisters when they walked by the tree.

The male mockingbird flew at their heads, squawking and scolding, glancing his sharp claws off their shoulders. The mockingbirds even chased away the dog and the cat if either one wandered up to the house from the barn.

As the oldest child, Joe solved the problem. He bravely claimed the privilege of wearing an old army helmet and going out the door first to distract the mockingbird. His younger brothers and sisters watched at the kitchen window. When they saw the mockingbird attacking Joe's helmet, they knew it was safe to run out and play.

© David Ponton

- The mockingbird can imitate (mock) as many as 30 different bird songs. It can even imitate the sound of a chirping cricket, a barking dog, and the back-up signal of a truck.

Mockingbird *(Mimus polyglottus)*

Size and color: About the size of a robin, mockingbirds are gray and black with 2 white bands on the wings and white patches under the wings.

Food: insects, fruit, berries, and spiders.

Habitat: shrubs, trees, streamside thickets, brushy deserts.

Nest and eggs: Mockingbirds build a cup of twigs, leaves and string in dense shrubs, trees, or vines from 1 to 50 feet off the ground. 3–6 green or blue speckled eggs, 2 or 3 clutches each season.

Migration: Mockingbirds generally stay all year.

Special interest: The Mockingbird is the Texas state bird because it is "a fighter for the protection of his home, falling, if need be, in its defense, like any true Texan . . ."

Behavior:

- Moonlight nights often are filled with the mockingbird's singing. While most people enjoy this late night song, some people close their windows, only to find the mockingbird "sings down the chimney."

GLOSSARY

clutch: a group of eggs.

intruder: unwelcome stranger, invader.

- ☐ In the spring, the male mockingbird chooses an area among shrubs and trees where he sings and waits for a female mockingbird. The female mockingbird chooses the male that is the best singer. After a period of courtship, the pair builds a nest of twigs and leaves.

Cowboy Story

J. Frank Dobie relates a story of a Texas cowboy who was riding fence one day. The cowboy saw a rattlesnake coiled and buzzing. The cowboy stopped to see what made the rattlesnake so mad. The rattlesnake started to slide away when a mockingbird appeared and flew down the rattlesnake's back and struck at the rattlesnake's head. The rattlesnake coiled again and the cowboy saw that one of its eyes was out and bleeding. He watched the mockingbird continue its attack on the rattlesnake for over thirty minutes. Each time the rattlesnake would try to slide away, the mockingbird would attack its head. "Finally, the bird struck the other eye, held on and flapped vigorously, then darted up. The rattler coiled and began striking in all directions, then sank his fangs into his own body . . . The mockingbird flew up into the bushes and began to sing. I put the rattler out of his misery with a large stone."

Adapted from *Rattlesnakes,*
by J. Frank Dobie

Once the nest is built, the male mockingbird becomes a fierce protector. He attacks any bird, animal, or person that comes into his territory. His loud singing and the flashing of wings let **intruders** know they are not welcome.

Western Kingbird

Annie and her father walked down the dirt road after supper, just to see what they could see. It felt good to get away from the house full of noisy relatives.

"Annie, why was the rhubarb pie under your bed?" Father asked. Annie gasped and looked down.

"Because I love rhubarb pie so much . . ." Annie said. "Oh, look! There's a kingbird on that pasture fence, see?" A yellowish bird sat proudly on a post, a dragonfly held firmly in its beak.

"Don't you just love kingbirds?" Annie pointed, "And the queen birds and the little prince birds and princess birds too?"

"Annie, it's not like a royal family. The female is still called a kingbird and the babies are called young kingbirds," Father said.

"Oh," Annie said. They walked a little further.

"Nice try at changing the subject," Father said. "About the rhubarb pie under your bed . . ."

"Well, I just had to hide it somewhere! You know how it is when the cousins come to visit. I only had one little piece of pie. I was afraid Leo and Ted and Dean and Dale and Bessie Fay would eat it all up. It was Cousin Bonnie's idea."

"You do have a lot of cousins," Father said sternly, placing his hand on Annie's shoulder. "Maybe we better hurry home and see if they've left us any."

© Rob Curtis/VIREO

The kingbird watches for insects from its perch on a fence wire or post. When it spots a bug flying by, the kingbird flies out to catch it, then returns to its perch. The kingbird may catch more than one insect at a time during its hovering flight.

Western Kingbird *(Tyrannus verticalis)*

Size and color: a bit smaller than a robin, gray head and chest with a yellow belly and black tail with white edges.

Habitat: open areas like pastures, fields, in cottonwoods along streams, and in urban areas, often near tall man-made structures.

Food: insects, most caught in the air.

Nest and eggs: often in cottonwood trees or on utility poles, windmills, or parts of buildings. The nest is made of twigs and grasses, lined with hair, string, cloth, or cotton. 2–5 creamy white eggs with blotches of brown.

Migration: Kingbirds spend the winter in Mexico and Central America.

Call/voice: *kip kip*

Behavior:

- The male kingbird sometimes makes a wild courtship flight to impress a female. The male flies high up into the air, then tumbles down, twisting and flipping like a crashing airplane. Before he reaches the ground, he will fly up and do it over again. If other male kingbirds are around, they too will often perform this tumble flight.

GLOSSARY

tumble flight: the bird flies up high and then tumbles back down like it is falling; done to impress another bird.

The Western Kingbird is a strong defender of its home. It chases away bigger birds like hawks and owls. The kingbird raises the red crown of feathers on the top of its head and bravely flies at the big birds, making a buzzing sound and snapping its bill.

"I wish I had beautiful feathers like that grackle," Cindy said. She watched the noisy birds strut across the grass in the park.

"I don't really like grackles. They remind me of pirates," Jaime replied. "Maybe it's the way they walk. They kind of swagger around like they think they're so important."

"But they're so pretty," Cindy said. "Sometimes their feathers look almost purple."

"Maybe *you* think they're pretty. But I bet you didn't know grackles eat other birds' eggs," Jaime said.

Cindy thought for a minute.

"Well, so do you," Cindy said. "So there!"

© J. Hoffman/VIREO

Grackles spend most of their time in flocks looking for food. They look for food on the ground, or by wading in shallow water, or searching trees for eggs and hatchlings in other birds' nests. Flocks of grackles often come into towns at night to roost in parks.

Great-tailed Grackle *(Quiscalus mexicanus)*

Size and color: larger than a robin, the male is black with glossy purple, greenish or bronze feathers, a wedge-shaped tail, and yellow eyes. The female is smaller and shorter tailed than the male. Male grackles are glossier than the dull, brownish females.

Habitat: farms, towns, parks, groves.

Food: omnivorous: variety of seeds, especially cracked corn, insects, salamanders, eggs of other birds, small fish, berries, whatever is available.

Nest and eggs: a cup shape made of sturdy twigs, stems, grass and even mud, usually in evergreens. 3–4 eggs, pale bluish or greenish with brown blotches.

Migration: Some northern flocks migrate to warmer areas in the winter. Some flocks stay in the same area all year.

Call/voice: loud piercing sounds, high pitched squealing and creaking, grating ratchety coarse noises.

GLOSSARY

omnivorous: will eat all kinds of foods.

Great-tailed Grackles spread their tails and wings and sing loud squeaking, grating songs while bobbing and bowing. Male grackles put on this show to attract a female.

What Does Little Birdie Say?

What does little birdie say,
in her nest at peep of day?
"Let me fly," says little birdie,
"Mother, let me fly away."
"Birdie, rest a little longer,
Till the little wings are stronger."
So she rests a little longer,
Then she flies away.

What does little baby say,
In her bed at peep of day?
Baby says, like little birdie,
"Let me rise and fly away."
"Baby, sleep a little longer,
Till the little limbs are stronger."
If she sleeps a little longer,
Baby, too, shall fly away.

—Alfred Lord Tennyson

Hummingbird

"Mom, what are you cooking?" Jaime asked as he watched his mother stir the pot of sweet smelling liquid.

"I'm making nectar for the hummingbirds," Mother said. "Most of the garden flowers have died and the hummingbirds need something to eat. This is sugar and water. I'll fill the feeder with it."

Jaime looked into the pot.

"Grandmother always put red food coloring in her hummingbird food. Did you forget that?" Jaime asked.

"I remember her doing that. Grandmother knew hummingbirds are attracted to the color red." Mother turned the heat off under the pot and waited for it to cool. "But now we know red dye is bad for the birds. The food coloring can make the birds sick."

"We don't need the red dye anyway," Jaime said. " The hummingbirds can't miss those big red plastic flowers on the feeder."

- At night, when the temperature drops, some hummingbirds have a special way of resting. In this resting state of **torpor**, the little bird's body temperature drops, requiring less energy. The bird's heartbeat slows from its normal rate of 1000 beats a minute to about 150 beats per minute.

© H.P. Smith/VIREO

Hummingbird Feeder Recipe

Mix together 1 cup white sugar and 4 cups water. Mix thoroughly. Many people heat the mixture to boiling to sterilize the water and help the sugar dissolve. Let the mixture cool before filling the feeder.

Tips

1. Use bottled water or boil the water.
2. Do NOT use food coloring.
3. Do NOT use honey (it can ferment).
4. Do NOT use sugar substitutes.
5. Clean feeders weekly with hot water but NO soap.
6. Empty the feeder completely and clean it with hot water before refilling.

Hummingbirds are fiercely territorial. Both the males and females choose high perches for watching over and guarding their territories.

Rufous Hummingbird *(Selasphorus rufus)*

Size and color: smaller than a sparrow, this medium sized (3 to 4 inches long) hummingbird has a rusty red color on its head and back. The male has a bright metallic red chin. The female is greener with red spots on her chin.

Food: nectar from red columbines, currants, manzanita, paintbrush, honeysuckle and other flowering plants, also small insects, sometimes tree sap.

Nest and eggs: a tiny nest (1 inch high, 1 inch deep) in bushes or trees. Nests are made out of plant fibers held together with spider webs. Lichen and bark are added to camouflage the nest. 2 eggs about the size of peas.

Migration: Migrates during the spring and summer as far north as Alaska, returning to the southwest United States and Mexico during the winter.

Call/voice: a low *chip chip chip*.

- Hummingbirds eat nectar, a sugary juice made by flowers. The hummingbird sticks its long bill into the flower to suck out the nectar. The flower's pollen falls onto the bill, head, or chin of the bird. The hummingbird then carries and spreads this sticky pollen to the next flower. This allows the plants to pollinate and create another generation of flowers.

- The hummingbird is the only bird able to fly backwards.

Other names:
Chupamirto dorada (Spanish). Hummingbirds are called "winged jewels."

Behavior:
- The hummingbird flutters its wings over 50 times a second as it hovers over its food source.

- Hummingbirds feed their babies by regurgitation. After eating, the mother sticks her long bill down the throat of the baby and regurgitates the food into the baby's crop.

GLOSSARY

camouflage: hiding by blending in with the surroundings.

nectar: the sweet liquid in a flower.

pollen: fine dust-like grain made by flowers for fertilization.

pollinate: to fertilize a plant.

regurgitate: to bring up from the crop or stomach, to vomit.

territorial: will defend its home area.

torpor: a time when a bird remains motionless to save energy during cold weather.

House Finch

© Joe Fuhrman/VIREO

The male House Finch is the colorful one. He may be pale yellow to bright red, depending on the colors of his food. Female House Finches prefer the reddest males.

"Oh Mom, what's the point? The House Finches just eat it right up. They're the most boring birds in the world," Jaime complained. Jaime's chore was to keep food in the bird feeders and the feeders were empty . . . again.

"Oh, I love to see them," Mother said. "When I was a little girl, we spent a year down in Mexico. I didn't know anybody and I was very lonesome, so Daddy bought me a House Finch in a cage. I named him Pequeño, and he was so tame he would perch on my hand and take a seed from my fingers. He was my best friend.

"Before we moved back here I set him free. And now—I know it's silly because it has been so many years—but now every time I see a House Finch I think maybe, just maybe, it's Pequeño.

"Okay, you win," Jaime said. "Here Pequeño. I'm coming . . . again."

House Finch *(Carpodacus mexicanus)*

Size and color: About the size of sparrows, the males are streaked brown, with red on the top of the head, above the eye, on the back, and shoulders. The female is brown.

Habitat: Finches live near humans or areas such as dry desert, grasslands, low elevation forests.

Food: seeds, buds, berries, fruit.

Nest and eggs: House Finches build a cup-shaped nest of grass in thick brush, in ivy on buildings, on ledges, street lamps, and places with overhead cover that protects the nest from rain. 4–5 pale blue eggs spotted with black.

Migration: Mostly year round residents, some flocks may move to a lower elevation in winter.

Call/voice: hoarse warble by the male. Like many birds, the males do most of their singing right after sunrise and right before sunset.

Behavior:

- House Finches are legally captured in Mexico and sold as pet birds.

- This finch often nests near buildings and is fond of feeding stations. This may account for the House Finch being North America's most familiar bird.

- When two House Finches are courting, one will gently peck at the other's beak. The female flutters her wings as if begging for food from the male.

☐ Once a young House Finch leaves the nest it will never return, even if it was disturbed from the nest and is unable to fly.

Goldfinch

"Mom, will you help me catch a bird? I want one of those bright yellow and black ones," Cindy flopped on the couch looking hopeless, "but they won't go under my box."

"You're trying to catch a goldfinch in a box?" asked her mother.

"Yeah, a cardboard box. I propped the box up with a stick, and I have a long string on the stick that goes over to the hedge. I'm hiding behind the hedge so I can pull the string and the box will fall down and trap the bird. Joe showed me how to do it. And I put breadcrumbs under the box so the bird will hop in under there, but no goldfinches ever come."

"Maybe goldfinches don't like bread crumbs? What do you see them eating?"

"They all go to that thistle feeder Jaime hung up." Cindy's legs dangled over the edge of the couch and she kicked her feet in frustration.

"Maybe you should put some thistle seed under your box and a goldfinch will come to eat it."

Cindy jumped up. "Hey! Good idea!" She ran outside. Tim, her brother, had been listening to the conversation.

"She still won't catch a goldfinch," Tim said to Mother.

"I know," Mother said. "But she won't get hurt trying and it keeps her out of trouble for a while."

© Tom and Pat Leeson

Finches like seeds still attached to a tree or stem. You will see finches clinging to a sunflower to eat the seeds. In contrast, sparrows would rather eat seeds fallen on the ground.

American Goldfinch *(Carduelis tristis)*

Size and color: smaller than a robin. During the breeding season, the male is lemon yellow with a black cap, black wings, and a white rump. The male and female look alike the rest of the year: olive yellow turning to light yellow, with dark wings and tail.

Habitat: streamside, woodland edges, orchards, suburban areas, open fields.

Food: seeds such as thistle, birch, alder, cedar, elm, weed seeds and insects. Goldfinches are particularly fond of thistle seeds in bird feeders.

Nest and eggs: The nest is built in July or August by the female goldfinch. She uses bark strips, weeds, vines, and grass. She strengthens the rim of the nest with spider silk and caterpillar webs. The nest is lined with thistle down, milkweed or cattail fluff. 4–6 bluish white eggs.

Migration: Some birds that have spent the summer in the northern states will migrate to the southwest for the winter. Some do not migrate if there is plenty of food.

- ☐ The beautiful colors of the male goldfinch come only after a spring molt. In the fall, the male molts again, turning back into a drab bird, the same color as the female.

American Goldfinch

Call/voice: canary-like song, *per-chic-o-ree*. Goldfinches have an undulating flight and often call or sing while flying.

Other names: Wild Canary

Behavior:

- During the mating season, you may see a female goldfinch crouch down and flutter her wings. She will give a begging call, just like a chick, and the male will bring her a seed. The female goldfinch is a drab color that helps her blend in with her surroundings while nesting.

- The goldfinch weaves her nest so tightly sometimes it can hold water. After a storm people have found baby goldfinches drowned by the water in the nest. Usually the mother goldfinch sits on the nest and spreads her wings like an umbrella, protecting the nest from gathering water.

GLOSSARY

molt: to shed old feathers and grow new ones.

undulating: up and down motion like a wave in the ocean.

House Sparrow

Be Like the Bird

Be like the bird, who
halting in his flight
On limb too slight
Feels it give way beneath him,
Yet sings,
Knowing he hath wings.

—Victor Hugo

House Sparrow *(Passer domesticus)*

Habitat: around houses and buildings in cities and on farms. They never live in the wilderness.

Food: seeds and insects.

Nest and eggs: a blob shaped nest made of grass and twigs, lined with feathers and built in small spaces in house eaves, drain pipes, and birdhouses. Will often take over other birds' nests. 3–6 whitish eggs with brown or gray spots on the larger end.

Migration: Does not migrate.

Call/voice: *cheep, chip, chirrup.*

Behavior:

- The House Sparrow is always found where people live or work. Its scientific name *domesticus* means "belonging to a household." House Sparrows often use their old nests as roosts in the winter. The nests can become very dirty and infested with fleas and mites.

☐ How do birds get into stores? Birds flying through the rafters *may be* House Sparrows that have learned to hover in front of electric eye sensors to open the doors.

Size and color: House Sparrows are chunky small birds. The wings and back are streaked brown and the chest is gray. They have a black bib and white cheeks.

☐ The House Sparrow was brought to North America and released in New York in 1851. The House Sparrow originally lived in Europe and North Africa, but it has survived here quite well and spread all across the United States

© Joe Fuhrman/VIREO

Nighthawk

Tim and Grandpa sat on the back porch eating peanut butter sandwiches. They chewed and looked west, where big dark clouds smothered the sun and darkened the sky.

"Darn it! I wanted to go swimming. No fair I can't go swimming the first day of summer vacation," Tim complained.

"Sorry, Sport," Grandpa said. "I'm glad for the rain. We sure need it."

"Look, the storm hawks are already out swooping around," Tim pointed to the dark sky. "I always see them flying around right before a storm."

"Those are really nighthawks, Sport. But I like your name for them. When I was your age, I thought those hawks pulled the storm clouds in. I was almost grown before I figured out the reason nighthawks fly around right before a storm. I believe it's because all the night-flying bugs come out."

© R. Curtis/VIREO

The nighthawk rests quietly on a branch or on the ground during the day. In the early morning, and again at evening, the nighthawk is out swooping around, scooping up flying insects with its wide mouth open.

Lesser Nighthawk *(Chordeiles acutipennis)*

Size and color: larger than a robin, mottled brown, buff, and black, with white crescents on the wings and throat.

Habitat: warm southern lowlands, deserts, dry washes, and riparian areas.

Food: night-flying insects of all kinds, especially flying ants.

Nest and eggs: bare flat pebbly ground, gravel streambeds, or on flat roofs. 2 eggs, clay colored, whitish, spotted with gray.

Migration: They fly to Mexico for the winter. A few nighthawks stay in their summer home for the winter in a state of torpor.

Call/voice: toad-like trill or soft whinnying. Nighthawks are generally silent.

- The mouth of the baby nighthawk is wide like a butterfly net. The mouth has hair-like bristle feathers that help sweep insects into it.

Behavior:

- Nighthawks respond to extreme temperatures in interesting ways. On very hot days, the nighthawk flaps a membrane in its throat to help cool off. During very cold times, the nighthawk stops moving to conserve energy. This state of torpor is like hibernation, but not as deep.

- The nighthawk is attracted to swarms of insects gathered around water or lights. You are likely to see this bird in towns around streetlights.

- Some nighthawks eat small rocks. The rocks stay in the bird's stomach and help grind the hard shells of the insects they eat.

- Nighthawks nest on the ground or on flat roofs. Sometimes the mother nighthawk will roll her eggs into the shade on very hot days. When the temperature drops, she will move them back to their original nest.

GLOSSARY

hibernation: a resting time of lowered activity.

riparian: near water like streams, ponds, rivers, lakes.

torpor: a time when a bird moves very little to save energy during cold weather.

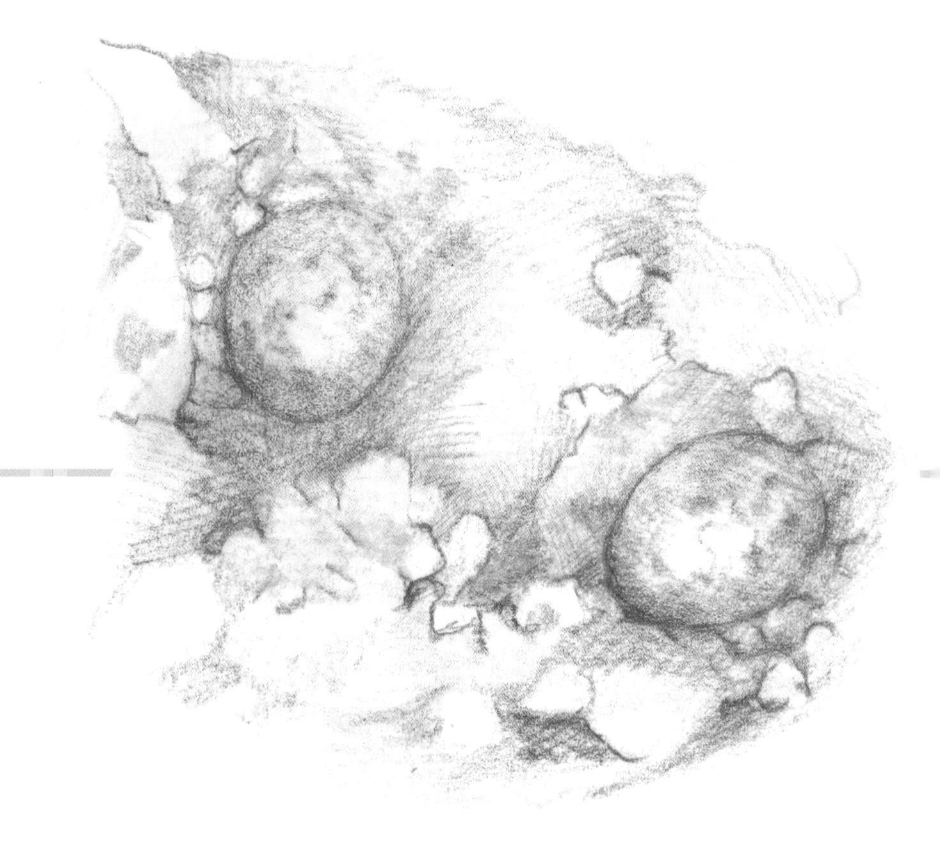

Mourning Dove

"The wind blew the bird nest down!" Cindy cried. She squatted to look closely at the two white eggs broken on the ground.

"Some birds build such flimsy nests," Tim said. "That's surely a Mourning Dove's nest. We're always finding their eggs rolled out and fallen to the ground. Maybe that's why some doves nest on the ground to begin with."

"But it's so sad," Cindy said. The dried liquid from one egg held the flattened shape of a very small bird. Cindy could make out the shape of its head and legs.

"Oh, don't worry," Tim said to comfort his little sister. "They'll build another nest soon and lay more eggs. There's never any shortage of Mourning Doves."

© Rick and Nora Bowers/VIREO

The Mourning Dove is one of the most **abundant** birds in the United States. Dove hunting is a popular sport for about two million people. This bird's name comes from its **mournful** song, a moaning *cooah, coo, coo, coo.*

Mourning Dove *(Zenaida macroura)*

Size and color: a long gray-brown bird, bigger than a robin. Has a small round head, plump body, and a long, pointed tail. The eye has a bluish ring of bare skin. The tail is edged with a black and white stripe. Short legs.

Food: Doves eat seeds off the ground. Doves love fields of corn, sunflower, wheat, and other seeds or grains. They also eat insects, fruits, nuts, acorns, and pine seeds. Sometimes eats snails in the spring before and during egg laying.

Nest and eggs: Usually nest in trees but will also nest on the ground. Doves build a nest of loosely fastened twigs and leaves. 2 white eggs, multiple broods, sometimes up to 7 broods a year.

Migration: Doves in the northern United States will migrate to the southwest and Mexico to avoid snow and low temperatures. If winters are mild, doves will stay year round.

Call/voice: a low soft, *cooah, coo, coo, coo.*

Behavior:

- Mourning Dove babies are fed crop milk or "pigeon milk" by their parents for their first three days. Crop milk is very nutritious and is regurgitated by both parents into the hatchlings' mouths.

- Mourning Doves burst away on whistling wings when surprised or frightened.

- Mourning Doves keep their bills in the water when drinking, taking steady sips. Other birds must tip their heads up so that water runs down their throats.

GLOSSARY

abundant: many, lots of something.

broods: a group of baby birds.

crop: a pouch inside a bird's throat for holding food.

hatchlings: baby birds too young to leave the nest.

mournful: sad.

pigeon milk: a liquid produced by a parent bird and fed to the hatchlings.

regurgitate: to bring up from the crop or stomach, to vomit.

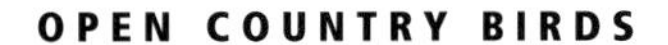

Quail

"Tell me a story, Mamma!" Tim begged as he climbed in bed.

"Alright," Mamma said. "This is a true story. It happened when I was a little girl. My brother Russell let me go with him when he drove across the fields looking for deer. One day Russell said, 'There's a quail nest that I've been watching. The eggs must be about to hatch.' He drove up to a clump of grass and we got out of the jeep. The mother quail was fluttering around the nest and making loud cries.

'Something's wrong!' Russell shouted. We ran toward the nest.

A huge snake stretched across the nest.

'Oh no!' Russell said. He picked up a stick and pinned the snake's head to the ground. Using his pocketknife, he cut off the snake's head. I saw the humps on the body of the snake and I knew right away what had happened to the quail's eggs.

Working quickly, Russell cut the snake's skin open to expose the newly swallowed eggs. He gently lifted the eggs to the grass and he picked at the cracked shells. The digestive fluid in the snake's belly had begun to dissolve the shells. Baby quail began to move. The baby birds were still alive!"

Gambel's Quail *(Callipepla gambelii)*

Size and color: The size of a robin, the male has the long topknot and a red helmet, a black face, gray-blue back, and wings with white stripes. The female is similar but not as colorful, with a shorter topknot.

Food: seeds, leaves, berries, cactus fruit, mesquite pods, insects.

Habitat: desert mesquite, canyons, especially around water which the covey visits each morning and evening. Some coveys have adapted to the suburbs, alfalfa fields, and vineyards

Nest and eggs: Quail nest on a shallow spot scratched on the ground, under a bush. The nest is large enough for 10–12 brown spotted eggs and is lined with grass. Coyotes and snakes eat many eggs. Quail sometimes use an old thrasher or roadrunner nest. Sometimes two females lay eggs in the same nest.

Migration: Quail are year round residents, though they may move down in elevation during the winter.

Call/voice: *whee-whit* or *ka-wha-whee-ah*, alarm call is *quit quit*

Behavior:

- Quail fly only as a last resort, to escape danger. If you ever startle a covey of quail, you know how the sudden explosion of birds and whirring wings at your feet stops your heart with fear and surprise. A quail reaches top flying speed very quickly, but cannot fly very far.

- Quail range far and wide until nesting season. Then they must choose a nesting site near water. Newly hatched quail babies must walk to water the first day they are hatched. Fortunately the young are highly precocial, able to walk and look for food within hours of hatching.

© Tom and Pat Leeson

GLOSSARY

covey: a flock of quail.

precocial: baby birds that are covered with down and able to leave the nest soon after hatching.

sentinel: one who watches for danger.

- ☐ A flock of quail feeds on the ground as one male perches in the top of a bush as a sentinel. When the sentinel quail sees danger, he sounds an alarm. The flock obeys the sentinel's warning and scatters in every direction.

Greater Roadrunner

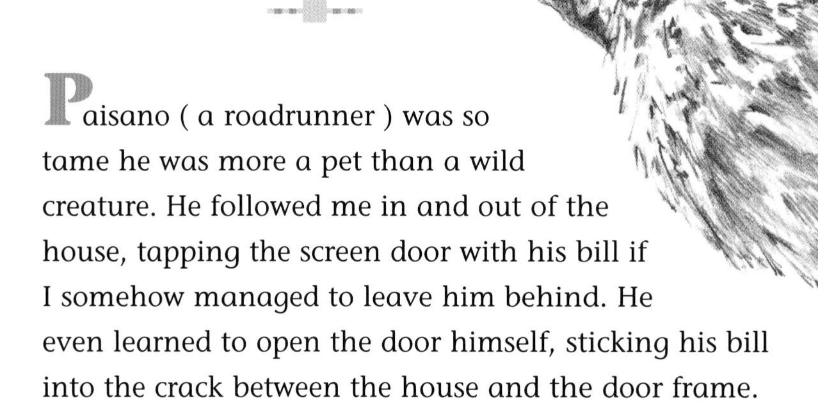

Paisano (a roadrunner) was so tame he was more a pet than a wild creature. He followed me in and out of the house, tapping the screen door with his bill if I somehow managed to leave him behind. He even learned to open the door himself, sticking his bill into the crack between the house and the door frame.

On a day in late November it rained hard all afternoon, a cold, dense rain that soaked the earth and turned the leaves on the ground dark brown. The autumnal monsoons had arrived. Paisano came into the living room and sat by the fire, refusing to budge. I let him be, understanding that somehow I'd domesticated him.

—To learn more about Paisano,
read *Paisano the Roadrunner*
by Jennifer Owings Dewey, 2002

Greater Roadrunner *(Geococcyx californianus)*

Size and color: bigger than a house cat, streaked brown and white, slender body, sturdy legs, bushy crest on male's head. The long tail is quite moveable.

Habitat: deserts, grasslands, brushy areas, some suburban areas.

Food: omnivorous: snakes, lizards, insects, scorpions, small rodents (mice, rats, bats), small birds (hummingbirds), cactus fruit.

Nest and eggs: a cup of sticks, lined with leaves and grass built in cactus, thickets or small trees. 2–6 white eggs.

Migration: year round residents. A layer of down under the outer feathers keeps the roadrunner warm in cold weather. They can expose a featherless patch of skin on the back to absorb heat from the sun on cold days.

Call/voice: dovelike coo's going down in pitch. Also a *perrp, perrp, perrp* sound and bill snapping.

Other names: Paisano (Spanish for countryman), Chaparral Cock, Ground Cuckoo, Snake Killer.

GLOSSARY

omnivorous: will eat all kinds of foods.

prey: animal killed for food.

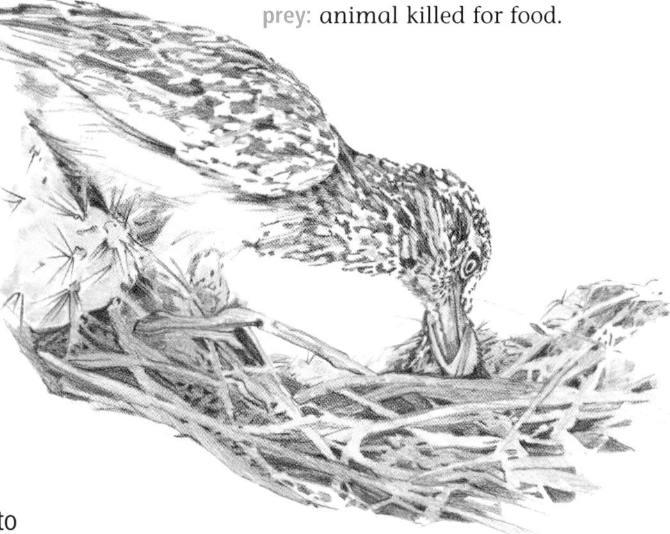

The mother roadrunner feeds her chick.

The roadrunner runs up to 18 miles per hour to catch its prey or to escape from danger. It seldom flies unless startled. The bird uses its long tail to steer itself, raising its tail to slow down or stop.

Greater Roadrunner

Folklore:

• Southwestern tribes associate the roadrunner with racing, courage, straightness, strength, and swiftness.

• The roadrunner, the State Bird of New Mexico, is seen as a tireless creature and is often shown on Zia Pueblo pottery.

• In Mexico, the roadrunner is a symbol of babies, much like the stork in other cultures.

Behavior:

• When you see a roadrunner's tracks, it is impossible to tell which way the bird was moving. The roadrunner track looks like an X, with two toes forward and two toes back.

• When roadrunners are courting, one bird may approach the other with a stick or a piece of grass in its bill. It will drop the stick or give it to the other bird as a gift.

☐ Roadrunners will work together to attack rattlesnakes. One bird distracts the snake while the other sneaks up and pins the snake's head. The birds then kill the snake by slamming its head against a rock.

Wetland Birds

Lakes, Streams, and Rivers

Canada Goose

One swallow does not make a summer, but one skein of geese, cleaving the murk of a March thaw, is the spring.

—Aldo Leopold, *Sand County Almanac*

Johnny slowly paddled his new red canoe across the lake. Cindy sat quietly in the front of the canoe, watching ducks fly up off the water when the boat floated near. Cindy spotted a long-necked mother goose on the shore of the lake, leading her babies through the puddles. Six small yellow **goslings** waddled behind their mother.

"Will the goslings be yellow when they're grown?" Cindy asked.

"No, Canada Geese are always brown and black," Johnny told her. "Brown feathers will replace their yellow down before winter."

"Let's go closer. Maybe I can catch one of the goslings and hold it. They look so soft," Cindy said.

"They are soft, but that father goose will never let you come close." Johnny stopped the canoe, "He's giving us the warning sign right now."

Cindy watched the big **gander** pump his head up and down his long neck and cry a loud "ahonk." The goslings huddled close to the mother goose, watching the canoe float across the lake.

Hatchlings often leave the nest the very first day out of the egg. Baby birds covered with down and able to walk right after hatching are called **precocial**. After all the eggs have hatched, the goose family walks, or swims, to an open grassy shore where there is plenty of grass to eat. Geese escape **predators** by going into the nearby water.

© Tom and Pat Leeson

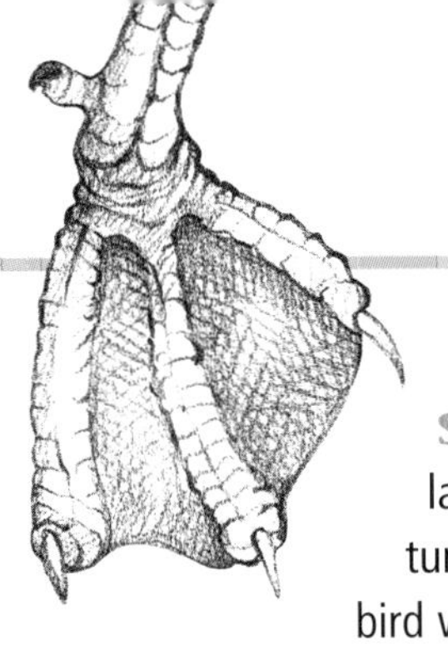

Canada Goose *(Branta canadensis)*

Size and color: a large (as big as a turkey) brown and white bird with white cheek patches and a black head and neck.

Habitat: lakes, ponds, marshes, grain fields.

Food: aquatic plants, short grass, grain, small aquatic animals.

Nest and eggs: nest built on the ground near open water, made of cattail leaves or grasses. 5 white eggs, 1 brood a year. The

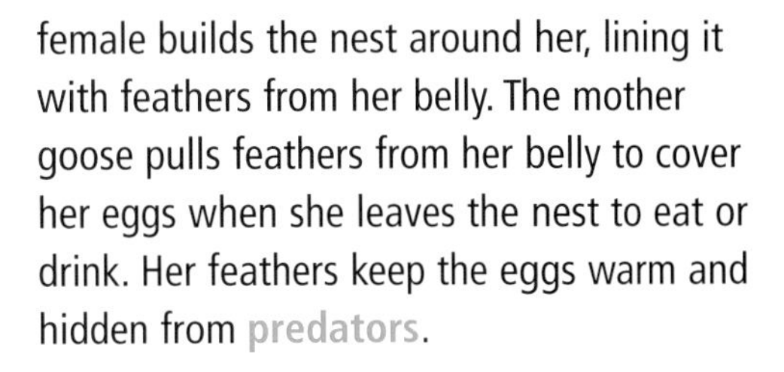

female builds the nest around her, lining it with feathers from her belly. The mother goose pulls feathers from her belly to cover her eggs when she leaves the nest to eat or drink. Her feathers keep the eggs warm and hidden from predators.

Migration: many migrate to the southern states during the winter and to Canada in the summer. Some flocks stay year round in urban areas and bird refuges.

Call/voice: The male goose calls *ahonk*, while the female's call is *hink*. The geese often call in a duet, alternating calls.

GLOSSARY

gander: a male goose.

goslings: baby geese.

incubating: warming the eggs while they develop.

molt: to shed old feathers and grow new ones.

precocial: baby birds that are covered with down and able to leave the nest soon after hatching.

predators: animals that kill for food.

- The gander (male) guards the female while she is on her nest incubating the eggs. The male's job is important and the female cannot successfully nest after the death of her mate. The male is fierce when defending his family. He will give a loud call and pump his head up and down before attacking anything that threatens his family.

Canada Goose

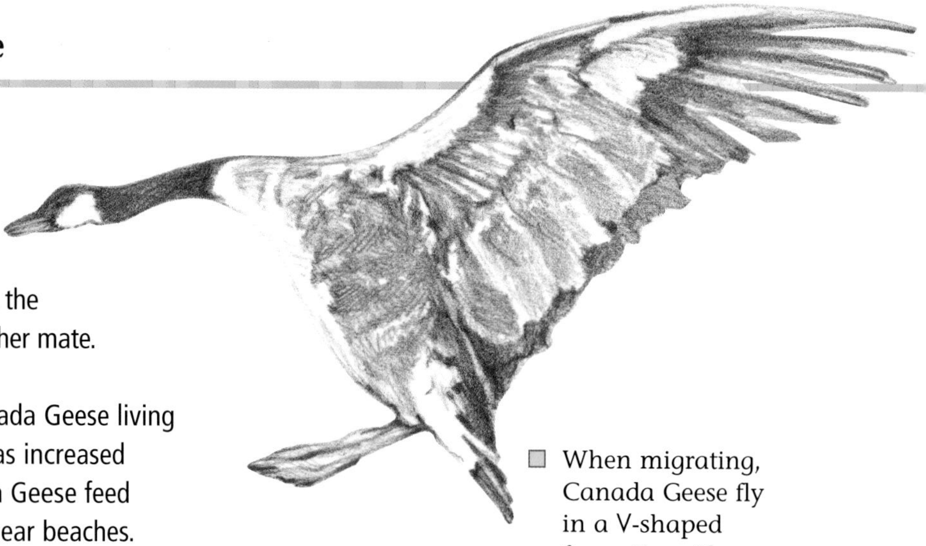

Behavior:

- Canada Geese mate for life. When one goose dies, the survivor will find another mate.

- The number of Canada Geese living in cities and towns has increased recently. Now Canada Geese feed on golf courses and near beaches. Fireworks, relocation, repellants, and herding dogs have all been used to discourage geese from occupying and damaging croplands and golf courses.

- Adult geese go through their molt while raising their babies. None of the goose family can fly while they are molting. By late summer all the members of the family are able to fly.

When migrating, Canada Geese fly in a V-shaped formation. They can fly for 600 miles or more without stopping to rest.

"Mamma, Mamma! Baby ducks are falling out of the tree!" Cindy jumped up and down yelling. She stepped back from the edge of the pond and watched a fluffy baby duck drop into the water and swim after the mother duck. Cindy's mother came running.

"Where? What tree?" her mother asked. "I've always wanted to see that."

Cindy pointed to a dead tree that hung over the water.

"How did baby ducks get up in the tree? They're too little to fly," Cindy said.

"They were born in a hole in that tree. When the eggs hatch, the babies climb out of the tree nest and jump down to the water," Mother whispered. "They start swimming and looking for food."

Cindy and her mother stood quietly and watched while the mother duck swam in a circle under the nest tree and called to her babies. Soon another baby duck poked its head out of a hole in the tree trunk and jumped into the water.

"I'm going to go jump out of that tree too," Cindy said.

"No. You'll scare them. They need to be with their mother and learn to find food. You stay here with me and we'll watch."

Wood Duck *(Aix sponsa)*

Size and color: A small duck, the male has an iridescent green and purple head and a white throat. He has a striking orange bill and reddish eye.

Habitat: marshes, streams, wetlands, beaver ponds, farm ponds, rivers.

Food: omnivore: seeds, fruit, acorns, insects.

Nest and eggs: Wood Ducks nest in cavities, but they cannot make the cavity. They prefer to nest over or near water. The nest is lined with down from the female's breast. 8–14 white or tan eggs.

Migration: Many Wood Ducks migrate to the Gulf of Mexico. Some are permanent residents and do not migrate.

Call/voice: high intensity *hauk*.

Behavior:

- Wood Duck ducklings leave their nest within a day of hatching. First, the mother duck looks out the nest hole to make sure it is safe for the babies. If it is safe, the mother flies from the nest to the ground or the water below the nest. The mother duck then calls her babies: *kuk . . . kuk . . . kuk.* The ducklings come to the mother duck's calls by climbing up with their clawed feet to the entrance hole in the tree. The babies jump out of the nest hole down to the mother. This can be a long way down and sometimes the ducklings are stunned. They recover quickly and join their mother. The mother duck leads the family, walking or swimming, to the nearest food.
- While Wood Ducks do not often dive beneath the water for food, they will dive down to escape predators.

GLOSSARY

dabble: to get food in the water with the bill without diving.

down: small fluffy feathers that keep a bird warm.

omnivore: an animal that will eat anything.

predator: animal that kills for food.

© Gene and Jason Stone/ Leeson Photography

Wood Ducks are dabbling ducks. Dabbling is a way of finding food on the surface of the water. Dabblers do not dive for food. This male Wood Duck is about to take flight.

Merganser

Mother opened the gift box and placed two stuffed birds on the mantel above the fireplace. One was a beautiful ring-necked pheasant. The other bird was some sort of funny-looking duck.

"That's about the ugliest looking bird I've ever seen," Tim said. "Where did it come from?"

"Your uncle sent those birds to us as a special gift," replied Mother.

"I guess he wanted to get rid of that ugly one," Tim scoffed.

"I don't know what you're talking about," Mother frowned. "I think they are both beautiful."

"Just look at it!" said Tim, moving closer. "It has funny looking feathers sticking up on its head and lots of little teeth. Funniest looking duck I've ever seen! What in the world is it?"

"It's a Merganser hen. And if you don't stop laughing at it, I'll make liver and onions again tonight for supper."

This female merganser dips its head under the water to spot fish. When the merganser sees a fish, it dives and swims under water after the fish, using its webbed feet to speed itself along. The merganser can stay under water for two minutes. The duck's bill has little sawteeth that help it hold its slippery **prey**.

Common Merganser *(Mergus merganser)*

Size and color: This large diving duck is almost as big as a goose. The male has an iridescent green head, white breast, and a black back. The female has a rusty head with a spiny crest, white breast, and gray wings.

Habitat: lakes with trees around them, rivers.

Food: Mostly fish. Also mussels, shrimps, insects.

Nest and eggs: prefers tree cavities near water, but will also use crevices in rocks and holes under tree roots. The female plucks soft down from her breast to line the nest. The spot where the down is plucked from her breast becomes the warm brood patch that keeps the eggs warm. 6–17 buff or yellowish eggs.

- Baby mergansers are precocial and leave the nest to swim and feed themselves within 2 days of hatching. Sometimes a tired baby merganser will crawl onto the swimming mother's back to rest. The mother duck does not stay with her brood all summer. She leaves the ducklings on their own within a month or two, even before they can fly.

Migration: Mergansers may flock together when they migrate south for the winter. They travel at night, following rivers. Many stay close to their summer breeding home during the winter.

Call/voice: *gruk gruk gruk* or *cro cro cro* when female calls young from the nest.

Other names: Sawbill, Fish Duck, Sheldrake. "Merganser" is Latin for "Diving Goose."

Behavior: Migrating flocks of mergansers work together to find food. They form a line in the water and drive the fish ahead of them. Mergansers patter across the water before flying into the air. They look like they are running across the top of the water right before they lift off.

GLOSSARY

brood: a group of baby birds all from the same batch of eggs.

brood patch: a warm place on a bird's belly where the feathers have fallen out and the skin is thick and full of blood. This patch is used to keep the eggs or hatchlings warm.

patter: the running motion of a water bird as it takes off from the water.

precocial: baby birds that are covered with down and able to leave the nest soon after hatching.

prey: animal killed for food.

Four ducks on a Pond

Four ducks on a pond,
A grass-bank beyond,
A blue sky of spring,
White clouds on the wing—
What a little thing
To remember for years!
To remember with tears!

—William Allingham

Red-winged Blackbird

Red-winged Blackbirds, like all birds, spend time preening themselves every day. They run their bills down each feather, smoothing and oiling all the parts into place. They take baths in fresh water and dust themselves in the dirt. Grooming feathers on a regular basis is an important job for every bird. A bird will die if its feathers are so damaged that it cannot fly.

Red-winged Blackbirds use ants to get rid of tiny bugs that injure their feathers. Mites are small insects that crawl into the birds' feathers. The mites hide between the barbules and eat the feathers. Baths don't get rid of mites, but ants do.

Ants secrete formic acid. When ants crawl around on birds, the formic acid is rubbed off on the feathers. This acid is a natural pesticide that mites cannot tolerate. The formic acid left by the ants either kills the mites or makes them leave.

At least 200 species of birds use ants to rid their feathers of mites. Some birds crush the ants and then rub them through their feathers. Other birds place ants among their feathers and let the ants crawl around. Some birds flop around on anthills and let the ants crawl up into their wings. Later, the birds will pick the ants out with their bills.

If there are no ants, birds will use other acidic substitutes: cigarette butts, termites, beetles, snails, chokecherries, and walnut meats. One family put mothballs in their garden to discourage the rabbits. They were surprised to see birds rubbing the mothballs on their feathers. The clever birds had found yet another way to get rid of the pesky mites.

Red-winged Blackbird *(Agelaius phoeniceus)*

Size and color: a glossy black bird, about the size of a robin, the male with bright red shoulder patches.

Habitat: marshes, grasslands, along irrigation ditches, wherever cattails grow.

Food: seeds, grains, insects, spiders.

Nest and eggs: woven of grass by the female, suspended from a stick or reed, often over water. 3–5 pale blue eggs, marked with black or brown zigzags.

Migration: Red-winged Blackbirds migrate from northern states to southern states for the winter.

Call/voice: *konk-la-ree, oak-a-lee, conqueree.*

GLOSSARY

barbules: the hair-like pieces of a feather.

epaulets: shoulder patches.

harem: a group of females all with the same mate.

mob: a group of birds chasing a predator away.

molt: to shed old feathers and grow new ones.

predator: animal who kills for food.

polygamous: having more than one female mate.

☐ Male Red-winged Blackbirds are polygamous, with two to nine mates. All the females in one "harem" nest close by each other. By nesting nearby, the females can protect each other's nests. The male watches for danger and switches calls to sound an alarm. One type of call warns of danger from above, such as a hawk or owl. A different call warns of danger from the ground, such as a human walking nearby. If a predator enters the Red-winged area, the males mob together to chase it away.

Behavior:

• The female is inconspicuous (dull colored). There is a good reason for the female to avoid being noticed. She spends most of her time among the reeds, building a nest or raising the young. It is easier for her to hide from predators when she can blend into her surroundings.

• Male Red-winged Blackbirds perform "song flights" over their territories, gliding slowly and singing. The male sings from his perch with his tail and wings spread, the shoulder epaulets flared out.

• One big enemy of Red-winged Blackbirds is the Marsh Wren. The Marsh Wren will destroy the eggs and kill the hatchlings of the Red-winged Blackbird. Raccoons, mink, and Black-billed Magpies also attack Red-winged Blackbirds' nests.

• Red-winged Blackbirds gather in flocks of millions of birds when the breeding season is over and their molt is complete. These flocks may take the shape of long columns in the sky that stretch for miles. Sometimes these birds flock with grackles, starlings, and cowbirds.

☐ Only the male Red-winged Blackbirds have the bright red epaulets (shoulder patches) that give the birds their name. The male shows his red patches to show he is claiming his territory. He covers up the red patches in times of danger when he does not want to be noticed.

Great Blue Heron

"Granddad, I see a great big bird real early every morning at the pond. It flies away into the mist when it sees me and I never can get a good look at it. What kind of bird is it?" Tim swirled a forkful of pancake in the syrup on his plate.

"Might be a heron, son. Does it have long legs?" Granddad poured himself another cup of coffee.

"Really long legs. They dangle behind when the bird flies. You want to go see if it's still there now?"

"Sure," Granddad said, standing up. "But remember, Tim, we have to walk slowly and we can't make any noise. No talking."

Tim and Granddad stopped walking the instant they spotted the big bird. They kept very still and watched. The bird stood in the shallows near the bank, its long legs like sticks, its neck bent like a snake. The strange bird had a long bill. In truth, everything about the bird looked long.

"Great Blue Heron," Granddad whispered.

The bird's bill shot forward as it grabbed something from the muddy bank of the pond. When the heron raised its head, Tim saw a frog wriggling in its yellow bill.

Tim couldn't stand it.

"Wow!" he exclaimed, and stepped forward. The heron lifted its wings and took off, drifting into the trees at the far end of the pond.

Great Blue Heron *(Ardea herodias)*

Size and color: When its neck is stretched up, it is as tall as a fire hydrant; a long-legged, long-necked blue gray bird. Herons can appear squat when all folded down. It has a black stripe above the eye and a long black plume of feathers trailing back from the head.

Habitat: marshes, beaches, riverbanks, swamps. They sometimes hunt in grasslands.

Food: mostly fish, but also frogs, lizards, birds and small mammals like voles.

Nest and eggs: The nest is made of sticks and lined with pine needles or grass, built in trees, bushes, or on the ground near water. Herons prefer islands and swamps. 3–5 dull pale blue eggs.

Migration: Herons migrate alone or in groups from summer breeding grounds to the southern coast of the United States and to the Caribbean, Central America, and South America. When migrating, herons fly both day and night.

Call/voice: loud *Frawnk.* They also make a loud bill snapping sound when defending an area.

© David Ponton

- Herons walk slowly through shallow water hunting for food. They often hunt at night because their eyes are adapted for night vision. When a heron spots a fish, it shoots its head and neck forward rapidly, catching the fish in its beak.

Other names: Crank.

Behavior:

• A mated pair of herons may each rapidly clap the tips of their bills at each other.

• Herons will hunt for food from floating objects like log rafts.

• Herons usually hold their necks in an S shape while flying. When they are courting, hunting, or defending their territory, they stretch their necks out.

• Great Blue Herons nest in colonies in hard-to-get-to places such as islands. Herons search for places where their nests will be safe from snakes and raccoons.

Sandhill Crane

The evening sky was dark with birds. The gray cranes flew with their long necks stretched out and their black skinny legs trailing behind. The air filled with the rolling trumpet sound of their calls. After feeding all day in the fields, the cranes gathered at the river to spend the night. The flock settled on a small island in the middle of the river. On the island, the birds were safe from hungry animals. The members of the many crane families found each other with hoarse creaking calls and settled in for the night.

Several cranes in each flock stood like sentries, their heads held high, watching for danger. Spring was coming to the southwest. Soon the cranes would be leaving on their long journey north.

© Tom and Pat Leeson

The Sandhill Crane is a winter visitor to the southwest. Cranes look for food in farm fields during the day. The cranes leave the fields at dusk to find a spot with shallow water or an island surrounded by deeper water. The cranes spend the night in these wide-open spaces, where predators, such as coyotes, will be easy to spot.

Sandhill Crane *(Grus canadensis)*

Size and color: The crane is a tall bird with a long neck and long legs. Adults are mostly mouse gray with a red patch of skin on the head. Young birds are reddish brown.

Habitat: prairies, farm fields, around marshes and river valleys.

Food: will eat most anything: grain, acorns, berries, small rodents, frogs, and insects.

Nest and eggs: they build what looks like a "soggy haystack" on the ground, with a moat of water around the nest. They lay 2 eggs, but usually raise only one chick.

Migration: The southwest flocks return to the northern United States and Canada in the spring. Young cranes learn the way from their parents.

Call/voice: a low loud growly throat rattle, *k-r-r-r-oo.*

Other names: The early settlers called the Sandhill Crane the "Preacher Bird." Young cranes are called colts.

☐ For young cranes, dancing is a bit like our "dating." Young cranes will dance with several cranes before settling down with a life-long mate.

Sandhill Crane

Lore:

- Cranes are honored in many cultures, especially Japan, where cranes symbolize long happy marriages.

- The expression of "craning" ones neck comes from the way a crane moves its head and neck.

Behavior:

- A group of cranes flying up in a spiral pattern is called a "kettle."

- Sandhill Cranes have razor-sharp toe claws.

- The crane's bill is long, strong, and straight: the perfect shape for picking through grass and poking down into mud.

☐ Sandhill Cranes dance, especially the "teenage" ones. They bow, spread their wings, dip down, and leap into the air. Sometimes a crane will pick up a stick and throw it in the air as part of the dance.

☐ Sandhill Cranes stay in large flocks (except during nesting season). Sandhill Cranes like to be in voice contact with others in the flock at all times. Fortunately, the voice of the crane can be heard over three miles. The crane's loud call is the result of its extra long windpipe. The crane's windpipe (trachea) is almost twice as long as its throat. The extra length is coiled behind its breastbone, sort of like a cinnamon roll.

Appendix

General Information about Birds

Size and Color: The first things you notice about a bird are its size and color. For many birds, the males' color is different from the color of the females. The males are usually the bright colors and the females are the dull colors. Young birds are often less colorful too, growing into their adult colors as they mature.

Habitat: (where the bird lives): Birds need the same things you need: food, water, and a safe place to rest and raise their families. Each type of bird lives where it can find food. Goldfinches eat seeds; bluebirds eat insects; herons eat fish; hawks eat meat. Geese eat grass.

Food: Habitat and food go together. If you know what a bird eats, you know where to look for the bird.

Nest and Eggs: Birds build nests and raise their babies in the spring and summer. Each kind of bird builds its own special type of nest.

Migration: When the weather turns cold, a bird still needs to eat. The seeds may be covered with snow. Many insects freeze and die. The ponds ice over. If a bird's food is too hard to find in the winter, the bird will fly to a warmer area with more food.

Some birds are "resident" in the southwest, like the chickadees and roadrunners. Resident birds do not migrate. Some birds, like the Sandhill Cranes, are winter visitors to the southwest and return to northern areas in the spring. Some birds, like hummingbirds, fly far south during the winter and return to the southwest in the spring.

Call/voice: Birds have their own language. They communicate with each other with calls, sounds, and postures. Some bird sounds mean "Here I am, look at me." Other calls mean "Watch out" or "Danger." A young bird calls "Feed me." A parent bird makes a "Stay away" sound if you come too close to its nest. It is very hard to write bird language in English. We have included the most common call that each bird makes.

Behavior: Birds are amazing. Woodpeckers drill holes with their beaks. Hummingbirds fly backwards. Orioles weave tight nests. Vultures eat rotting flesh. Blackbirds rub ants through their feathers. Mockingbirds can imitate the calls of other birds. If you know some of the ways birds behave, you will never be bored.

How to attract birds to your yard

Food

- Put seeds, such as sunflower seeds, in a feeder or on the ground.
- Finches and other small birds like to eat thistle seed out of a thistle feeder.
- Peanut butter and suet attract birds in winter.
- Slices of apple or orange placed on the ground attract birds that do not eat seeds.
- Hang up a hummingbird feeder.
- Throw old cereal into the yard.

Water

- Birds love to bathe, and of course they need water to stay alive. Keep a birdbath full of fresh water. If you don't have a birdbath, use a shallow pan from the kitchen.

Plants, trees, and bushes

- Trees and bushes give shelter and nesting sites so birds feel safe.
- Plants, trees, and shrubs provide food for birds. Insects live in the plants and the birds eat them. Birds love the seed that plants make.

Index